SELF-ASSEMBLY IN SUPRAMOLECULAR SYSTEMS

Monographs in Supramolecular Chemistry

Series Editor: J. Fraser Stoddart, FRS
University of California at Los Angeles, USA

This series has been designed to reveal the challenges, rewards, fascination and excitement in this new branch of molecular science to a wide audience and to popularize it among the scientific community at large.

No. 1 Calixarenes
By C. David Gutsche, Washington University, St. Louis, USA

No. 2 Cyclophanes
By François Diederich, University of California at Los Angeles, USA

No. 3 Crown Ethers and Cryptands
By George W. Gokel, University of Miami, USA

No. 4 Container Molecules and Their Guests
By Donald J. Cram and Jane M. Cram, University of California at Los Angeles, USA

No. 5 Membranes and Molecular Assemblies: The Synkinetic Approach
By Jürgen-Hinrich Fuhrhop and Jürgen Köning, Freie Universität Berlin, Germany

No. 6 Calixarenes Revisited
By C. David Gutsche, Texas Christian University, Fort Worth, USA

No. 7 Self-assembly in Supramolecular Systems
By Leonard F. Lindoy, University of Sydney, New South Wales, Australia, and Ian M. Atkinson, James Cook University, Townsville, Queensland, Australia

How to obtain future titles on publication

A standing order plan is available for this series. A standing order will bring delivery of each new volume immediately upon publication. For further information please contact:

Sales and Customer Care Department, Royal Society of Chemistry, Thomas Graham House, Science Park, Milton Road, Cambridge CB4 0WF, UK
Telephone: +44 (0)1223 432360; Fax: +44 (0)1223 423429; E-mail: sales@rsc.org

Monographs in
Supramolecular
Chemistry
Series Editor
J. Fraser Stoddart, FRS

Self-assembly in Supramolecular Systems

Leonard F. Lindoy
University of Sydney, New South Wales, Australia

Ian M. Atkinson
James Cook University, Townsville, Queensland, Australia

RS•C
ROYAL SOCIETY OF CHEMISTRY

ISBN 0-85404-512-0

A catalogue record for this book is available from the British Library

Published by The Royal Society of Chemistry,
Thomas Graham House, Science Park, Milton Road, Cambridge CB4 0WF, UK

For further information see our web site at www.rsc.org

Typeset by Wyvern 21, Bristol
Printed in Great Britain by Cambridge University Press, Cambridge, UK

Preface

Molecular self-assembly is a widespread phenomenon in both chemistry and bio-chemistry. Yet it was not until the rise of supramolecular chemistry that attention has been increasingly given to the *designed* self-assembly of a variety of synthetic molecules and ions. While success in this enterprise has to a large extent reflected the learning of Nature's lessons, it has also been aided by an increased awareness of the latent steric and electronic information implanted in individual molecular components. It seems that there are few areas in contemporary chemistry where human creativity can be so readily expressed.

While generally not yet approaching the sophistication of biological assemblies in either form or function, synthetic systems of increasing subtlety and – as we have attempted to show – very considerable aesthetic appeal, have been created. The story so far is an exciting one, with real promises for the future.

In this monograph, we survey highlights of the progress made in the creation of *discrete* synthetic assemblies. It is our hope that the work will form a foundation (and indeed a motivation) for new workers in the area as well as also being useful to experienced supramolecular chemists. It may also aid workers in the biological area to see Nature's assemblies in a new light. Further, the approach employed has been designed to provide readable background material for use in senior undergraduate and graduate courses in the supramolecular area.

The authors thank their families in Townsville and Sydney for their forbearance during the preparation of the manuscript. We also express our gratitude to members of the Lindoy research group at the University of Sydney for both their proof-reading and helpful comments. Finally, we express our gratitude to Professor, the Lord Lewis FRS, and Fellows of Robinson College, Cambridge, where much of the text was written.

L.F. Lindoy

University of Sydney,
Sydney, New South Wales

and

I.M. Atkinson

James Cook University,
Townsville, Queensland

Contents

CHAPTER 1

Self-assembly: What Does it Mean?

1.1 Introduction

Supramolecular chemistry – broadly the chemistry of *multicomponent* molecular assemblies in which the component structural units are typically held together by a variety of weaker (non-covalent) interactions – has developed rapidly over recent years.[1] 'Typically' is used since, in a considerable number of systems, metal–donor bonds – often essentially covalent in nature – have also been employed to 'stitch' together organic components into larger assemblies. Such metal-linked assemblies will be treated as part of the supramolecular realm in the present work (although not employed here, perhaps 'supermolecular' is a better term for this category).

With the development of supramolecular chemistry, there has been a concomitant shift in the mind-set of chemists working in the area. This has involved a change in focus from single molecules, often constructed step by step *via* the formation of direct covalent linkages, towards molecular assemblies, with their usual (see exception above) non-covalent weak intermolecular contacts. This change in focus is nicely encapsulated in Lehn's description of supramolecular chemistry as 'the designed chemistry of the intermolecular bond'.[2]

As a consequence of the intense interest in the field, a very large number of synthetic supramolecular systems have now been synthesised, with many of the (non-polymeric) systems ranging in size from around a nanometre or so to tens of

[1]   J.-M. Lehn, Nobel Lecture, *Angew. Chem., Int. Ed. Engl.*, 1988, **112**, 90; *Comprehensive Supramolecular Chemistry*, eds. J.L. Atwood, J.E.D. Davies, D.D. MacNicol and F. Vögtle, Pergamon, Oxford, 1996, vol. 11 and preceding volumes; *Monographs in Supramolecular Chemistry*, ed. J.F. Stoddart, Royal Society of Chemistry, Cambridge, see preceding volumes in the present series; J.-M. Lehn, *Supramolecular Chemistry: Concepts and Perspectives*, VCH, Weinheim, 1995; G.M. Whitesides, E.E. Simanek, J.P. Mathias, C.T. Seto, D.N. Chin, M. Mammen and D.M. Gordon, *Acc. Chem. Res.*, 1995, **28**, 37; I. Dance, 'Supramolecular Inorganic Chemistry', ch. 5, in *The Crystal as a Supramolecular Entity*, ed. G.R. Desiraju, 1996, Wiley, Chichester, UK, pp. 137–233; D. Philp and J.F. Stoddart, *Angew. Chem., Int. Ed. Engl.*, 1996, **35**, 1154; *Transition Metals in Supramolecular Chemistry*, ed. J.-P. Sauvage, Wiley, New York, 1999; F.T. Edelmann and I. Haiduc, *Supramolecular Organometallic Chemistry*, Wiley/VCH, New York, 1999.
[2]   J.-M. Lehn, *Angew. Chem., Int. Ed. Engl.*, 1990, **29**, 1304.

nanometres. Quite often, innovative design features have been required to achieve the desired structures – with the design and synthesis of individual systems often representing a very considerable intellectual and practical achievement. The field remains an exciting and fast moving one that continues to produce a range of new materials; many of which are endowed with aesthetically pleasing structures as well as unusual properties. The latter, for example, may include novel redoxactive, photoactive, conductive (including superconductive) and non-linear optical behaviour. Clearly, the area is one that continues to show considerable promise for underpinning the development of molecular scale components and devices, including opto-electronic devices. The promise of *useful* molecular devices remains a motivation for the continuing widespread interest in the field.

Much of the work in supramolecular chemistry has focused on molecular design for achieving complementarity between single molecule hosts and guests. Besides complementarity, recognition, self-assembly, preorganisation and even self-replication represent important 'key words' in the armoury of the supramolecular chemist. As a consequence, the practice of supramolecular chemistry tends to be a somewhat interdisciplinary activity, often requiring knowledge of a range of appropriate chemical, physical and biochemical procedures and techniques. Indeed, aspects of supramolecular chemistry now impinge on virtually all of the chemistry sub-disciplines.

Apart from the special case where metal ions are used as the 'glue', central to the supramolecular field is the use of a variety of weaker (non-covalent) interactions – including hydrogen bonding, π–π stacking, dipolar interactions, van der Waals forces and hydrophobic interactions – to hold molecular components together. These are the same forces that Nature uses to bind its molecular assemblies. Indeed, much of the activity in the area aims to mimic (but not necessarily copy directly) the way that Nature goes about things.[3]

Creativity and challenge are, by necessity, key ingredients in any effort to devise and synthesise totally synthetic molecular systems that function like biological systems. To achieve such an aim, the elements of molecular recognition, self-assembly and (ultimately) self-synthesis, all ubiquitous in biology, need to be mastered. Further, the product of such a synthesis should be capable of being functionally active if it is truly to match the behaviour of a natural system. The work discussed in this and subsequent chapters documents the progress made, across a broad front, towards this goal. While some quite beautiful examples of self-assembled synthetic systems have now been produced (very often in good yield and under mild conditions), in general there is still a long way to go before individual systems match the biological ones in both subtlety and function. Therein lies the challenge! Indeed, the entire synthetic supramolecular enterprise so far tends to be dominated by the interaction of relatively simple molecular components that are associated with a limited number of bonding contacts on forming the aggregated product. In contrast, for larger biological assemblies, such as DNA, the tobacco mosaic virus, the enzymes or the respiratory proteins, the respective components are of high molecular weight and are of a quite complex nature. Indeed, the resulting assemblies

[3]   E.C. Constable and D. Smith, *Chem. Br.,* 1995, 33.

typically incorporate hundreds, if not thousands, of intermolecular contacts. Many systems of this type are able to reassemble from their separated components; the amount of steric and electronic information stored in the latter, and which must be 'read out' during reassembly, is thus very large indeed. Such levels of information storage (and processing) have not yet been approached in the synthetic systems investigated so far. Inevitably, a move towards greater complexity will represent one direction for future development.

How might higher molecular weight assemblies be produced? One (of many possible) *modus operandi* would be to mimic Nature by stringing together complementary molecular units in predetermined sequences such that two matched strands are produced that will induce self-assembly over many tens, or even hundreds of nanometers of strand length. In such an approach, the characteristics of the final assembly would be set by the nature, positioning in the strand sequence, and frequency of incorporation of the individual complementary molecular units together with their relative 'cross-strand' orientations. So far, the use of such a 'modular unit' approach for the construction of larger synthetic assemblies has been little exploited.

Of course, the above suggestion ignores the problem of possible supramolecular functionality. Whereas the natural systems are invariably characterised by high functionality in terms of their biochemical roles, in contrast, the functionality of the majority of synthetic assemblies so far investigated has very often been either minimal or, indeed, absent altogether. The incorporation of designed functionality into supramolecular systems will thus undoubtedly continue to attract increased attention in future studies.

1.2   Self-assembly

In the present context, self-assembly may be defined as the process by which a supramolecular species forms spontaneously from its components. For the majority of synthetic systems it appears to be a beautifully simple convergent process, giving rise to the assembled target in a straightforward manner.[4]

It must be emphasised that self-assembly is very far from a unique feature of supramolecular systems – it is ubiquitous throughout life chemistry. Biological systems aside, self-assembly is also commonplace throughout chemistry. The growth of crystals,[5] the formation of liquid crystals,[6] the spontaneous generation of synthetic lipid bilayers,[7] the synthesis of metal co-ordination complexes,[8] and the alignment of molecules on existing surfaces[9] are but a few of the many manifestations of self-assembly in chemical systems.

4   D. Amabilino and J.F. Stoddart, *New Scientist*, 1994, 25; D.S. Lawrence, T. Jiang and M. Levett, *Chem. Rev.*, 1995, **95**, 2229.
5   *The Crystal as a Supramolecular Entity*, ed. G.R. Desiraju, 1996, Wiley, Chichester, UK.
6   P. Espinet, M.A. Esteruelas, L.A. Oro, J.L. Serrano and E. Sola, *Coord. Chem. Rev.*, 1992, **117**, 215; R. Bissell and N. Boden, *Chem. Br.*, 1995, 38.
7   P.F. Knowles and P.G. Stockley, *Chem. Br.*, 1995, 27.
8   J.P. Collin, P. Gavina, V. Heitz and J.-P. Sauvage, *Eur. J. Inorg. Chem.*, 1998, 1; E.C. Constable, *Chimica*, 1998, **52**, 533; C.J. Jones, *Chem. Soc. Rev.*, 1998, **27**, 289.
9   C.D. Bain and S.D. Evans, *Chem. Br.*, 1995, 46.

A distinctive feature of using weak, non-covalent forces, or for that matter metal–donor bonds, in molecular assemblies is that such interactions are normally readily reversible so that the final product is in thermodynamic equilibrium with its components (usually *via* its corresponding partially assembled intermediates). This leads to an additional property of most supramolecular systems: they have an in-built capacity for *error correction* not normally available to fully covalent systems. Such a property is clearly of major importance for natural systems with their multitude of intermolecular contacts. It is a factor that will assume increasing importance for the construction of the new (larger) synthetic systems mentioned previously – as both the number of intermolecular contacts present and overall structural complexity are increased.

Nevertheless, thermodynamic reversibility may prove a disadvantage in particular systems. A classic example[10] is provided by the synthesis of individual rotaxanes ('bead on a thread' compounds), discussed in detail in Chapter 4. After threading of the macrocyclic 'bead' on to an open-chain component by self-assembly, it has been found desirable to block the reverse (unthreading) pathway by subsequent covalent attachment of branched alkane groups to each end of the 'thread' so that any tendency for separation of the components is blocked. That is, the final covalent step is equivalent to tying a knot in each end of the 'thread' to stop the 'bead' from slipping off.

A related example of this type is given by the facile synthesis of catenanes – supramolecular compounds incorporating mechanically interlocked rings[10] (these are discussed in Chapters 5 and 6). An efficient procedure for synthesising these compounds involves an initial self-assembly process in which an open-chain component is threaded through a macrocyclic component, then orientated such that a subsequent 'template' ring-closing reaction results in a structure containing the required mechanically locked rings.

Both of the above examples correspond to assembly processes that involve non-covalent followed by covalent bond formation; of course, other sequences are also possible as is totally covalent self-assembly – strategies discussed by Lindsey as early as 1991.[11]

It needs to be noted that supramolecular systems may also form under kinetic rather than thermodynamic control. This situation will tend to be more likely for larger supramolecular assemblies incorporating many intermolecular contacts, especially when moderately rigid components are involved. It may also tend to occur when metal ions, and especially kinetically inert metal ions, are incorporated in the framework of the resulting supramolecular entity or when, for example, an intermediate product in the assembly process precipitates out of solution because of its low solubility.

There are some difficulties in discussing the concept of self-assembly in the present context. First there is a problem of nomenclature. The host–guest conven-

[10] P.R. Ashton, R.A. Bissell, D. Philp, N. Spencer and J.F. Stoddart, in *Supramolecular Chemistry*, eds. V. Balzani and L. De Cola, Kluwer, Dordrecht, 1992, pp. 1–16; R.E. Gillard, F.M. Raymo and J.F. Stoddart, *Chem. Eur. J.*, 1997, **3**, 1933.
[11] J.S. Lindsey, *New J. Chem.*, 1991, **15**, 153.

tion, largely derived from simple macrocyclic chemistry,[12] was defined at the outset as representing species with concave and convex binding sites, respectively.[13] The term is now also commonly used in relation to the assembly of supramolecular systems; however, host–guest suggests the presence of two or more 'unequal' partners. As such, this original connotation is not necessarily appropriate for describing the components of individual supramolecular systems. In view of this, although sometimes used within the limits of their original meanings, we have very often made no attempt to distinguish hosts from guests in our discussion of molecular assemblies.

Related to the above is the question of size. At what point does a host–guest complex become 'supramolecular'? Currently, at least in terms of common usage, the answer seems to lie *in the eye of the beholder*. We will maintain this tradition.

While a considerable amount of data is now available covering the thermodynamic aspects of assembly formation, in general, very little corresponding information is available concerning their detailed mechanisms of formation and dissociation. In particular, little insight exists into the co-operative nature of individual host–guest contacts in directing the course of the assembly processes. As an aid to efficient supramolecular assembly design, this is clearly an area requiring further attention.

1.3 Molecular Recognition

Like self-assembly, molecular recognition processes are found widely throughout both natural and synthetic systems. In particular, molecular recognition is, of course, of central importance in a range of biological and medical areas including, for example, fields as diverse as immunology, pharmacology and genetics. Likewise, it is clearly of fundamental importance to a number of chemical areas. These range from sensor and other analytical applications, through separation science, to aspects of catalysis. Molecular recognition is also crucial to organic templating effects which, in themselves, represent a specialised form of self-assembly process.[14]

As discussed more fully in Chapter 2, it is the degree of electronic and steric complementarity between host and guest that, in general, dictates the magnitude of any molecular recognition that occurs for a given supramolecular system. However, other subtleties may also influence recognition behaviour. When a discrete molecular assembly has a stoichiometry other than 1 : 1, the existence of more than one host–guest binding domain raises the prospect that co-operative binding may occur – akin to allosteric behaviour in natural systems.

The presence of chirality in host and guest will likewise affect the interaction between them. Chirality can perhaps be seen as a 'second order' source of stored structural information that is available for exploitation, often with dramatic effect, for achieving an additional type of host–guest recognition based on 'handedness'.

[12] L.F. Lindoy, *The Chemistry of Macrocyclic Ligand Complexes*, Cambridge University Press, Cambridge, 1989.
[13] D.J. Cram and J.M. Cram, *Acc. Chem. Res.*, 1978, **11**, 7.
[14] R. Hoss and F. Vögtle, *Angew. Chem., Int. Ed. Engl.*, 1994, **33**, 375; S. Anderson, H.L. Anderson and J.K.M. Sanders, *Acc. Chem. Res.*, 1993, **26**, 469.

The idea of host preorganisation, first proposed by Cram,[15] provides a means for rationalising much guest recognition behaviour as well as the (observed) relative binding strengths of many supramolecular complexes. Essentially, the *preorganisation effect* implies that the more closely the binding sites of a host molecule are arranged for binding to a guest, the larger will be the association constant for the corresponding host–guest complex. In such a case, there will thus be minimal change in the degrees of conformational freedom of the host on binding to the guest, especially if the host's backbone structure is rigid. As a consequence, this lower 'loss of disorder' of the host is expected to be manifested as a favourable entropic contribution to the overall free energy of host–guest complex formation.

It is instructive to consider the relationship between molecular preorganisation, recognition and self-assembly in relation to the formation of a supramolecular complex. Classically, all three will play a sequential role in complexation. Namely, appropriate *preorganisation* of the bonding sites in the host for receiving the guest thus predisposes the former for guest *recognition*. This, in turn, promotes spontaneous *self-assembly* of the required supramolecular entity. In part, the stability of the guest depends upon the degree of preorganisation of host with respect to guest since the forces acting in the recognition step will also, in essence, continue to act in the product after formation.

Besides the above, the overall stability of a supramolecular complex will clearly also depend upon both the number and the nature of the available binding sites in each component. As discussed in Chapter 2, solvation effects may often also play an important role in determining the strength of host–guest binding.

1.4   Scope of the Present Treatment

Owing to the enormous range of what can now be described as supramolecular chemistry, the present work will chiefly focus on *discrete* molecular assemblies. In particular, no attempt has been made to discuss the burgeoning field[16] of 'crystal engineering'. Similarly, unless of particular relevance to the topic under discussion, higher oligomeric and polymeric systems, including most metal cluster and related compounds, have also been excluded from the discussion.

In the next chapter, we will look in more detail at the nature of the weak intermolecular forces that act between the assembled components of supramolecular systems.

[15]   D.J. Cram, *Angew. Chem., Int. Ed. Engl.*, 1986, **25**, 1039; D.J. Cram, *Angew. Chem., Int. Ed. Engl.*, 1988, **27**, 1009; J.A. Bryant, C.B. Knobler and D.J. Cram, *J. Am. Chem. Soc.*, 1990, **112**, 1254; J.A. Bryant, J.L. Ericson and D.J. Cram, *J. Am. Chem. Soc.*, 1990, **112**, 1255.

[16]   G.R. Desiraju, *Crystal Engineering: The Design of Molecular Solids*, Elsevier, Amsterdam, 1989; C.B. Aakeroy and K.R. Seddon, *Chem. Soc. Rev.*, 1993, **22**, 397; S. Subramanian and M.J. Zaworotko, *Coord. Chem. Rev.*, 1994, **137**, 357; M.J. Zaworotko, *Chem. Soc. Rev.*, 1994, **23**, 283; J.C. McDonald and G.M. Whitesides, *Chem. Rev.*, 1994, **94**, 2383; A.D. Burrows, C.W. Chan, M.M. Chowdhry, J.E. McGrady and D.M.P. Mingos, *Chem. Soc. Rev.*, 1995, 329; G.D. Prestwich, *Acc. Chem. Res.*, 1996, **29**, 497; R. Bishop, *Chem. Soc. Rev.*, 1996, **25**, 311; C.B. Aakeroy, *Acta Crystallor., Sect. B*, 1997, **53**, 569; S.C. Zimmerman, *Science*, 1997, **276**, 543; D. Braga and F. Grepioni, *Comments Inorg. Chem.*, 1997, **19**,185.

CHAPTER 2

Intermolecular Interactions: The Glue of Supramolecular Chemistry

2.1 Introduction

In the past, synthetic chemists have largely focused on the *reaction* of molecules rather than on their *interaction*. However, at least for the supramolecular chemist, this no longer holds true. Increasingly, attention has been given to the formation of molecular assemblies that are held together by a range of relatively weak inter-molecular interactions, much as Nature holds its molecular assemblies together. These non-covalent interactions are often dominated by hydrogen bonding and, if aromatic components are present, by π–cloud interactions. Also, other weak forces (both attractive and/or repulsive) may act. These include dispersion, polarisation and charge-transfer interactions – combinations of which make up van der Waals forces. While the use of hydrogen bonding and π–π interactions have tended to receive most attention in the design of individual supramolecular systems, van der Waals considerations are often also of crucial importance.

Knowledge of the various types of non-covalent interactions involved in host–guest formation is, of course, of fundamental importance to the design of new host–guest systems; it is also necessary for understanding the structural interplays that control the assembly of known systems both in the solid state and in solution.

In this chapter, an overview of the respective intermolecular forces that may con-tribute to self-assembly processes is presented, with emphasis on those aspects that appear of relevance to a practising supramolecular chemist.

2.2 Ionic and Molecular Recognition

Ionic or molecular recognition by a host molecule will depend upon the degree of structural and electronic complementarity between host and guest.[1] Structural

---

[1]   G.C. Maitland, M. Rigby, E.B. Smith and W.A. Wakeham, *Intermolecular Forces: Their Origin and Determination*, Oxford University Press, Oxford, 1981.

complementarity is typically associated with the presence of a cavity or cleft in the host incorporating fixed or semi-fixed binding sites that are correctly aligned for binding to the guest. Electronic complementarity involves binding sites (or surface areas) on the host and guest that are compatible in terms of their electron density distributions. It may, for example, involve the choice of less specific hydrophobic environments in the host that complement the overall nature of the guest (see Section 2.9 for a discussion of hydrophobic interactions). Alternatively, it may be dominated by charge complementarity focused more narrowly on specific areas of the host and guest.

Of course, for many host–guest systems the distinction between structural and electronic effects is not necessarily clear cut. It is the subtle interplay between both the nature and shape of the respective potential energy surfaces of the two (or more) molecules in a host–guest system that will control both molecular recognition and binding strength.

2.3 The Basics

A *molecule* can be considered to be a group of atoms whose binding energy is much greater than kT at room temperature. At the other extreme, *van der Waals clusters* are loose assemblies of atoms or molecules whose binding energy is of the order of kT so that thermal collisions at room temperature can readily dissociate them (examples are noble gas clusters and aromatic moiety/noble gas complexes). Such clusters will thus normally only occur as discrete entities at low temperatures. Between these two categories lie *supramolecular (host–guest) complexes*. These are systems that have definable structures and stoichiometries and which are generally associated with measurable binding constants at room temperature.

In supramolecular assemblies the integrity of the individual component molecules normally remains largely intact; that is, the wave functions of the respective molecular components remain largely separate on complex formation. Overall, the net sum of all intermolecular forces binding such a discrete host–guest system is typically less than the strength of a single covalent bond. Namely, the total intermolecular interaction will rarely be greater than around 100 kJ mol^{-1} (while the weakest covalent bonds are on the order of 150 kJ mol^{-1}).

In broad terms, molecular forces may be considered to fall into two main classes:[1–3]

- short-range – these are mainly of the coulombic- and exchange-type that include covalent bonds and result from orbital overlap. They can be attractive or repulsive and in particular instances may represent the strongest forces present in a molecular system.
- long-range – these consist of those interactions that may be broadly characterised as being proportional to r^{-m} (r = intranuclear distance, m is a positive integer),

2   A.D. Buckingham, A.C. Legon and S.M. Robers, *Principles of Molecular Recognition*, Blackie Academic & Professional, Glasgow, 1993.
3   N.W. Alcock, *Bonding and Structure*, Ellis-Horwood, Chichester, 1990.

these include van der Waals, electrostatic (including hydrogen bonding) and π–π interactions. Long-range forces are those that contribute primarily towards supramolecular complexation.

2.4 Electrostatic Interactions

Electrostatic interactions between static molecular charges – such as found, for example, between polar molecules – tend to be relatively strong as well as direction dependent. As such, they are often of central importance in molecular recognition. They can be attractive or repulsive and are the simplest of the intermolecular forces to account for since their effects are additive.

The dielectric constant of the medium can have a large effect on the strength of particular electrostatic interactions. For example, two unit charges of opposite sign 5Å apart in a vacuum have an interaction energy of about -280 kJ mol^{-1}. However, this may be reduced by almost two orders of magnitude in polar media. It must be noted that the potential between ions is roughly proportional to r^{-1} when r is small but to r^{-2} when r is large[1,2] (r is the relevant interatomic distance).

2.5 Hydrogen Bonds

Hydrogen bonds can be considered to represent a special type of electrostatic interaction. The term hydrogen bond was coined in 1920 to help describe the internal structure of water. However, since that time the precise meaning of the term has been subject to change so that it is difficult to define it in a manner that will satisfy all. A simple working description is that it is an attractive interaction between a proton donor and a proton acceptor. In this vein, Pimentel and McClellan[4] in their classic text define a hydrogen bond as follows:

A hydrogen bond exists between a functional group A–H and an atom or a group of atoms B in the same or a different molecule when:
  (a) there is evidence of bond formation (association or chelation);
  (b) there is evidence that this new bond linking A–H and B specifically involves the hydrogen atom already bonded to A.

Both the donor (A) and acceptor (B) atoms have electronegative character, with the proton involved in the hydrogen bond being shared between the electron pairs on A and B.

Hydrogen bonds can vary in strength from being very weak to being the strongest (and most directing) of the intermolecular interactions. Spectroscopic as well as structural tools, including IR and NMR spectroscopy as well as neutron/X-ray diffraction,[5] have been widely employed to investigate hydrogen bonding; the range and diversity of hydrogen bonding types recognised has risen steadily with time.

4   G.C. Pimentel and A.L. McClellan, *The Hydrogen Bond*, Freeman, San Francisco, 1960.
5   I. Olovsson and P.G. Jönsson, in *The Hydrogen Bond: Recent Developments in Theory and Experiments*, P. Schuster, G. Zundel and C. Sandorfy, eds., North Holland, Amsterdam, 1976, vol 1I.

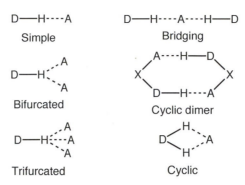

Figure 2.1 *Common arrangements of hydrogen bond types*

Common arrangements may be described as *simple* (this type is often bent rather than linear as shown), *bifurcated*, *trifurcated*, *bridging* or *cyclic* (Figure 2.1).

The most common hydrogen bond donor groups are C–H, N–H, O–H, S–H, P–H, F–H, Cl–H, Br–H and I–H while acceptor groups include N, O, P, S, F, Cl, Br and I as well as alkenes, alkynes, aromatic π-clouds and transition metals;[6] acceptor groups are characterised by being associated with significant areas of electron density.

Hydrogen bonds involving acceptor atoms that are more electronegative than carbon generally correspond to the strongest intermolecular forces found in supramolecular systems. For neutral species, bond strengths are typically of the order of 5–60 kJ mol^{-1}. Moreover, when one of the participants in the hydrogen bond is ionic, the strength of the interaction can rise to well above this range.[7]

It is only relatively recently that there has been wide recognition of a weaker (but important) category of hydrogen bond; namely, of type C–H⋯A (A = F, O, N, Cl, Br, I).[6,8,9] For example, in the case of C–H⋯N, there are a very large number of such contacts listed in the Cambridge Structural Database that are significantly shorter than the sum of the respective van der Waals radii (2.75 Å). Contraction of this type has been used as a criterion for the existence of hydrogen bonding.[10] In general, bonds of this type are promoted when the C–H hydrogens involved are relatively acidic.

The inherent directionality of hydrogen bonds makes them ideal for use in achieving complementarity in supramolecular systems. Most hydrogen bonds are bent and X-ray structural data have been drawn upon to correlate bond lengths with angles for a wide range of hydrogen bonds.[10] Systems incorporating weaker (longer) hydrogen bonds exhibit a greater tendency to be associated with bond angles showing a

6   C.B. Aakeröy and K.R. Seddon, *Chem. Soc. Rev.*, 1993, **22**, 397.
7   M. Moet-Ner, *J. Am. Chem. Soc.*, 1983, **105**, 4906; M. Moet-Ner, *J. Am. Chem. Soc.*, 1984, **106**, 1257.
8   G.R. Desiraju, *Acc. Chem. Res.*, 1991, **24**, 291; G.R. Desiraju and T. Steiner, *The Weak Hydrogen Bond in Structural Chemistry and Biology*, Oxford University Press, Oxford, 1999.
9   S. Subramanian and M.J. Zaworotko, *Coord. Chem. Rev.*, 1994, **137**, 357.
10 M. Mascal, *Chem. Commun.*, 1998, 303.

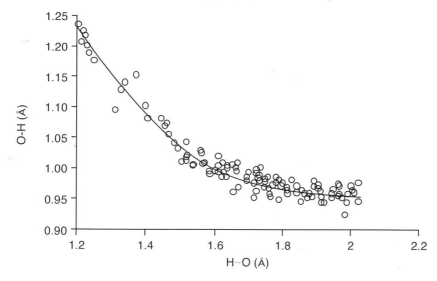

Figure 2.2 *O–H bond length as a function of O⋯H bond length for a series of accurately determined neutron structures. (Hydrogen bonds with O–H⋯O angles greater than 150° are omitted from the analysis.)*

greater deviation from linearity, with N–H⋯N bonds appearing to be more likely bent than O–H⋯O bonds. As might be anticipated, an inverse correlation between D–H and H⋯A lengths (D is the donor atom, A is the acceptor atom in the hydrogen bond) is observed[5,11,12] (see Figures 2.2 and 2.3).

The nature of hydrogen bonds can usually be more readily understood in terms of the shape of the different potential wells that may characterise individual hydrogen bond donor–acceptor systems. Possible potential functions are illustrated in Figure 2.4.[13]

Another type of hydrogen bond involves π-facial interactions of the type illustrated in Figure 2.5.[14-16] As an approximation, such arrangements can be considered to bear a relationship to both a classical donor–acceptor hydrogen bond as well as to a π–π bonded system (see below). Interactions of this type tend to be quite weak (1–5 kJ mol⁻¹). However, they often act in a co-operative manner with other intermolecular interactions such that, for example, they help to dictate a precise orientation within a given supramolecular architecture.

[11] R. Taylor and O. Kennard, *Acc. Chem. Res.*, 1984, **17**, 320.
[12] T. Steiner, *J. Chem. Soc., Chem. Commun.*, 1995, 1331.
[13] J. Emsley, *Chem. Soc. Rev.*, 1980, **9**, 91.
[14] H. Adams, F.J. Carver, C.A. Hunter and N.J. Osborne, *Chem. Commun.*, 1996, 2529.
[15] H. Adams, K.D.M. Harris, G.A. Hembury, C.A. Hunter, D. Livingstone and J.F. McCabe, *Chem. Commun.*, 1996, 2531.
[16] C.A. Hunter, *Chem. Soc. Rev.*, 1994, **23**, 101; C.A. Hunter, J. Singh and J.M. Thornton, *J. Mol. Biol.*, 1991, **218**, 837.

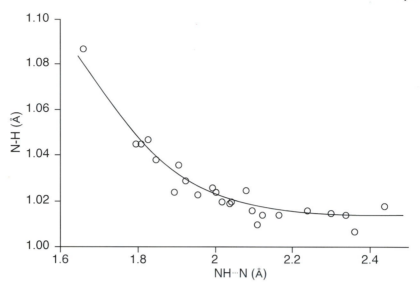

Figure 2.3 *N–H bond length as a function of N···H bond length for a series of accurately determined neutron structures.*[12] *(R < 0.06, only N–H···N bonds with no additional contacts are included in the analysis.)*

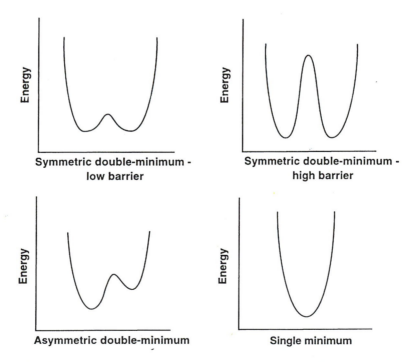

Figure 2.4 *The four potential functions that may be used to classify donor–acceptor hydrogen bonds*

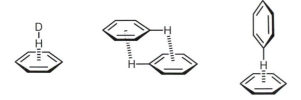

Figure 2.5 *Weak hydrogen bond π-facial interactions*

2.6 Van der Waals Interactions

As mentioned already, van der Waals interactions are a collective group of long range inductive and dispersive intermolecular forces. They act between molecules at distances generally larger than the sum of their electron clouds. Van der Waals forces again tend to be weak but their effects are additive; hence their total collective contribution to complex stability can be significant. The component forces can once again be grouped in various ways. Classical *inductive* forces are attractive and include permanent dipole–dipole and induced dipole–dipole interactions. The magnitude of these interactions varies as an inverse power of the distance between the interacting species.

Dispersion forces (sometimes called London forces) are quantum mechanical in nature and result from momentary fluctuations of the electron density within the electron cloud of molecules. Dispersion potentials vary with r^{-6} and are generally of quite small magnitude. However, since they are additive, their combined effect can again also be significant. As an example, while the dispersive force between two CH_2 groups separated by 5Å is only approximately –0.3 kJ mol^{-1}, for a long stretch of n parallel hydrocarbon chains, this results in an approximate dispersion energy of $-n1.7$ kJ mol^{-1}, a significant attraction. As with other electrostatic effects, the strength of inductive and dispersive interactions is very dependent upon the media. In general, inductive forces are attenuated in polar media. Dispersive interactions may, however, be enhanced under such conditions.

2.7 π–π Interactions

Interactions between π-systems have long been observed in crystal structures of aromatic molecules and, for example, play a role in stabilising DNA through vertical base-pair interactions. They are also involved in the intercalation of drugs into the grooves of DNA. Such interactions have also been exploited in many synthetic host–guest systems.[17] Previous rationales for the occurrence of π–π interactions have included explanations based on solvophobic,[18] electron donor–acceptor[19] as

[17] C.G. Claessens and J.F. Stoddart, *J. Phys. Org. Chem.*, 1997, **10**, 254.

[18] H.-J. Schneider, K. Philippi and J. Pohlmann, *Angew. Chem., Int. Ed. Engl.*, 1984, **23**, 908; A.R. Fersht, *Enzyme Structure and Mechanism*, Freeman, New York, 1985, pp. 293–310; J. Canceill, L. Lacombe and A. Collet, *J. Chem. Soc., Chem. Commun.*, 1987, 219.

[19] K. Morokuma, *Acc. Chem. Res.*, 1977, **10**, 294; R.L. Strong, *Intermolecular Forces*, B. Pullman, ed., D. Reidel, Dordrecht, 1981, pp. 217–232.

well as atomic models.[20] However, by themselves, none of these approaches is fully satisfactory for modelling the structures of π–π complexes.

While π–π attractions can be well modelled with high-level *ab initio* methods, which implicitly include quadrupolar, inductive and short-range interactions between adjacent π-systems, it has been difficult in the past to analyse such interactions successfully using simpler methods. However, in 1990, Hunter and Sanders[21] proposed a conceptually simple model for treating such interactions, which is based upon electrostatic and van der Waals forces. Their approach proved successful in accounting for many of the characteristics of such interactions. The treatment enables analysis of the 'offset' face-to-face and T-geometry arrangements for both neutral and polarised π-systems and, for example, has been successful in enabling observed molecular orientations to be understood. Further, the rules are useful for the prediction of the nature of π–π interactions and form a basis for the design of efficient host–guest systems incorporating such interactions.

The main feature of the model is that it considers the σ-framework and the π-electrons separately. It concludes that apparently net favourable π–π interactions are not due, in fact, to attractive electronic interactions between the two π-systems, but rather occur when the attractive interactions between the π-electrons and the σ-framework (namely, π–σ attractions) outweigh the unfavourable π–π repulsions that are present. Although van der Waals forces also contribute to the interaction energy, it is the electrostatic π–σ interactions that dictate the preferred geometry to be adopted.

In the simplest case, the model is based on the concept of an idealised π-system that consists of a positively charged σ framework (+1e) sandwiched between two regions of π-electron density ($2 \times -\frac{1}{2}e$). This situation is illustrated in Figure 2.6. Point charges are used to represent these regions, the distance between them being governed by the experimental value of the quadrupole moment of benzene.

Any polarisation of the system (for example, arising from the presence of substituents) is then accounted for by modifying the value of the point charges (these values are obtained from MO calculations on the non-π-bonded aromatic system). An interaction surface is then calculated for all geometries of interest – the atom–atom electrostatic interactions and van der Waals interactions are then calculated, their sum being the net potential energy of the π–π interaction for the given geometry. It is noted that the potential energy surface built up in this way does not include components arising from dispersive, inductive or short-range interactions. Nevertheless, at least qualitatively, the procedure has been successful in predicting the interactions between π-systems over a range of structures. The calculated interaction surface for two benzene rings is shown in Figure 2.7.[16]

Based on their calculations, Hunter and Sanders[21] have developed six simple rules that apply to π-stacked aromatic systems, with the last three applicable to polarised systems.

[20]   A.V. Muehldorf, D. Van Engen, J.C. Warner and A.D. Hamilton, *J. Am. Chem. Soc.*, 1988, **110**, 6561.

[21]   C.A. Hunter and J.K.M. Sanders, *J. Am. Chem. Soc.*, 1990, **112**, 5525.

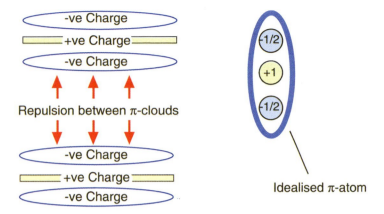

Figure 2.6 *Hunter and Sanders' model of a π-system (left).[21] A positively charged σ-framework is sandwiched between negatively charged π-clouds. Face-to-face interactions in this model are clearly repulsive. The diagram on the right represents an idealised atom that contributes one π-electron to a molecular system*

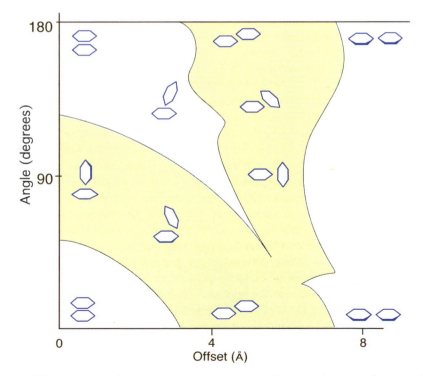

Figure 2.7 *Map of the electrostatic interaction between benzene rings as a function of their relative orientation. Attractive interactions are those that fall within the coloured region[16]*

(1) π–π Repulsion dominates in aligned face-to-face π-stacked geometries [see Figure 2.8(a)].
(2) π–σ Attraction dominates in edge-on (T) geometries [see Figure 2.8(b)].
(3) π–σ Attraction dominates in offset π-stacked geometries [see Figure 2.8(c)].
(4) For interactions between highly charged atoms, charge–charge interactions dominate.
(5) A favourable interaction with a neutral or weakly polarised site requires the following π-polarisation: (a) a π-deficient atom in a face-to-face geometry; (b) a π-deficient atom in the vertical T-group in the edge-on geometry; and (c) a π-rich atom in the horizontal T-group in the edge-on geometry.
(6) A favourable interaction with a neutral or weakly polarised site requires at least one of the following σ-polarisations: (a) a positively charged atom in a face-to-face geometry; (b) a positively charged atom in the vertical T-group in the edge-on geometry or (c) a negatively charged atom in the horizontal T-group in the edge-on geometry.

It is to be noted that reversing the polarisations listed in Rules 5 and 6 results in repulsion.

It is also worth emphasising the important (counter-intuitive) observation arising from Rule 5 that favourable face-to-face interactions can take place for π-deficient systems.

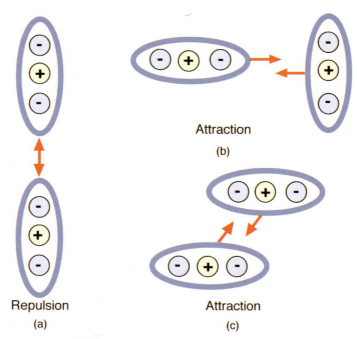

Figure 2.8 *Hunter and Sanders' model[21] of interactions between idealised π-atoms (Rules 1–3): (a) in a face-to-face interaction π-atom repulsion dominates; (b) whereas π-σ attraction dominates in a T interaction; (c) as well as in an offset π-stacked geometry*

2.8   Charge-transfer Interactions

Charge-transfer (CT) interactions, sometimes termed electron donor acceptor (EDA) interactions, result from the mixing of the ground and excited charge-separated states (that is, AB and $A^+ B^-$ states). These are very weak intermolecular forces. Charge-transfer complexes may form when good electron donors and acceptors lie in close proximity. This situation is typically associated with charge-transfer transitions in the UV–Vis spectral region. CT interactions have frequently been used to explain the stabilisation of interacting π-systems although, as discussed, the above Hunter and Sanders electrostatic model appears to present a better approach for interpreting structure–function relationships.[21]

Once again, solvation effects can strongly influence the strength of host–guest binding in such systems. In systematic studies using a range of aromatic-containing substituted cyclophanes as the hosts and a variety of neutral, 2,6-disubstituted naphthalene and *para*-disubstituted benzene derivatives as the guests, Diederich and co-workers[22] demonstrated that stable and highly-structured inclusion complexes form. The dominating host–guest interactions in these complexes are π–π stacking and edge-to-face aromatic–aromatic interactions. In organic solvents, EDA interactions were shown to control the *relative stabilities* of the respective cyclophane–arene inclusion complexes. As predicted, the more electron-deficient benzene and naphthalene derivatives form their most stable complexes with the electron-rich cyclophanes under these conditions.

In water, however, the above contribution from EDA interactions tends to be over-ridden by solvation effects. In this solvent, solvophobic forces (see below) can provide a very favourable contribution to the free energy of association of host and guest.

2.9   Hydrophobic Binding

Hydrophobic binding may be described as the association of non-polar regions belonging to individual host and guest molecules in aqueous or other protic media – reflecting the tendency of the solvent molecules to seek their most stable hydrogen-bonded environment. As a consequence, solvation effects are very important in mediating the strength of hydrophobic binding.[23] Even though the nature of a given hydrophobic interaction may be rather ill-defined, its existence may nevertheless have major implications for both host–guest formation and molecular recognition.

Hydrophobic binding plays a major role in both the folding and recognition behaviour of large biomolecules.[24] Thus, with many biopolymers, where recognisable regions of polar and non-polar residues exist, the tendency for the non-polar regions to associate and thereby exclude polar solvent molecules is well documented. Studies of protein folding and substrate recognition using molecular

[22]   F. Diederich, D.B. Smithrud, E.M. Sanford, T.B. Wyman, S.B. Ferguson, D.R. Carcanague, I. Chao and K.N. Houk, *Acta Chem. Scand.*, 1992, **46**, 205.
[23]   H.-J. Schneider, R. Kramer, S. Simova and U. Schneider, *J. Am. Chem. Soc.*, 1988, **110**, 6442.
[24]   W.L. Jorgensen and D.L. Severance, *J. Am. Chem. Soc.*, 1990, **112**, 4768.

dynamics simulations have in recent times greatly increased the understanding of these processes.[2]

Hydrophobic bonding is also common in supramolecular chemistry (the very large number of cases in which hydrophobic guests bind to the hydrophobic interior of a cyclodextran are well known examples of this type). Complex formation involving hydrophobic binding will, as mentioned above, be accompanied by partial or full desolvation of host and guest and, clearly, the strength of solvation of the separated entities will affect the stability of the resulting complex. Water is the most favourable solvent for inducing association between hydrophobic species but, conversely, any host–guest complexation component involving hydrogen bonding will be reduced in this solvent.

The magnitude of hydrophobic binding free energy has been estimated to be about 0.2 kJ mol^{-1} per Å^2 of host–guest contact.[25] This positive contribution will of course be moderated by the loss of entropy on binding, reflecting the loss of translational and rotational freedoms of the separated components.

3.0 Final Comments

While non-covalent bonding interactions may define and direct the self-assembly processes that leads to new supramolecular systems, it is important to note that in general such forces still continue to act once the respective systems are formed. As such, they will continue to govern any dynamic processes that occur within the self-assembled structure.

It also needs to be mentioned that while weak intermolecular interactions have been emphasised in this chapter, a given supramolecular structure may also contain stronger covalent linkages between its individual components. As discussed in Chapter 1, in many systems this involves metal ion co-ordination in which the metal links individual components within the overall structure.

[25] D.H. Williams and M.S. Westwell, *Chem. Soc. Rev.*, 1998, **27**, 57; L. Serrano, J.-L. Neira, J. Sancho and A.R. Fersht, *Nature (London)*, 1992, **356**, 453; K.A. Sharp, A. Nicholls, R. Friedman and B. Honig, *Biochemistry*, 1991, **30**, 9686.

CHAPTER 3

Hydrogen-bonded and π-Stacked Systems

3.1   Introduction

In this chapter the formation and properties of discrete hydrogen-bonded and π-stacked supramolecular entities are discussed. In many cases the bonding in these systems mimics that found in biological assemblies. Biological systems aside, the number of (discrete) artificial assemblies of this type now reported runs into hundreds, if not thousands, and collectively they represent a very large proportion of the synthetic supramolecular entities so far reported. While an attempt has been made to be representative in the present coverage, it has been, of course, not possible (or appropriate) to be comprehensive. For convenience of discussion, the categories of supramolecular species represented by the rotaxanes, the catenanes and assemblies incorporating metal ions as part of their superstructure are not covered in the present chapter (even though many of these species also contain hydrogen bonds and/or π-intermolecular interactions). Instead, each of these supramolecular categories is treated separately in subsequent chapters. As already mentioned in Chapters 1 and 2, much emphasis has been given to the use of directed hydrogen bonds to obtain molecular recognition in particular in host–guest systems.[1] In such cases, hydrogen bond formation can be viewed as reflecting the presence of 'programmed directional information' in both host and guest. On the other hand, and as discussed in Chapter 2, π–π interactions are also directional but in a less focused way;[2] they are also generally weaker than hydrogen bonds. In many supramolecular systems both types of interaction act in concert to produce the final structure. This, for example, is beautifully illustrated in Nature by the structure of DNA which assembles through hydrogen bonding between its base pairs and π-stacking between aromatic moieties in adjacent double helical strands.

While the role of hydrogen bonding and π-interactions are our principal concern in the examples to be discussed, clearly other influences of the type discussed

---

[1]   Y. Aoyama, in *Supramolecular Chemistry*, V. Balzani and L. De Cola, eds., Kluwer, Dordrecht, 1992, pp. 17–30.
[2]   T. Dahl, *Acta Chem. Scand.*, 1994, **48**, 95; C.A. Hunter, *Chem. Soc. Rev.*, 1994, **23**, 101.

previously (including dispersion forces as well as steric, coulombic, entropic, medium and chiral influences) will also affect, sometimes profoundly, the spontaneous self-assembly of a final structure.

3.2 Simple Hydrogen-bonded and/or π-Stacked Assemblies

3.2.1 Simple Host–Guest Assemblies

One of the simplest expressions of self-assembly is the 1 : 1 complexation of a host and guest. The investigation of such systems over many years has provided a cornerstone for the development of supramolecular chemistry involving larger molecular aggregates. It is appropriate that we look first at a few simple systems.

The synthetic hosts in such studies have typically incorporated multiple hydrogen bonding sites which are often strategically positioned in a cleft or cavity for binding to a chosen complementary guest.[3] Owing to competition for the hydrogen bonding sites, many of these systems have been demonstrated to form strong host–guest aggregates only when a non-polar solvent is employed.[4] Frequently, when aromatic groups are present, π-interactions of one type or another[5] also contribute to the overall stability of the complex generated.

Self-assembly, involving dimerisation of a single suitable monomer, corresponds to the simplest type of 'host–guest' complexation. Thus, the dithiourea derivative **1**,

1

containing *n*-butyl side chains, self-assembles to form a dimeric structure in which two molecules of **1** are orientated orthogonally in a 'head-to-tail fashion', with the four thiourea groups involved in a closed network of hydrogen bonds.[6] The dimer was demonstrated to form in deutero-chloroform using ^{1}H NMR and the structure was shown to persist in the solid state. NMR studies, vapour pressure osmometry

[3]   A.D. Hamilton, in *Advances in Supramolecular Chemistry*, G.W. Gokel, ed., Jai Press, Greenwich, CT, 1990, vol. 1, p. 1; J. Rebek, *Acc. Chem. Res.*, 1990, **23**, 399.
[4]   J.C. Adrian and C.S. Wilcoz, *J. Am. Chem. Soc.*, 1991, **113**, 678; D.H. Williams, J.P.L. Cox, A.J. Doig, M. Gardner, U. Gerhard, P.T. Kaye, A.R. Lal, I.A. Nicholls, C.J. Salter and R.C. Mitchell, *J. Am. Chem. Soc.*, 1991, **113**, 7020.
[5]   See, for example: K.R. Adam, I.M. Atkinson, R.L. Davis, L.F. Lindoy, M.S. Mahinay, B.J. McCool, B.W. Skelton and A.H. White, *Chem. Commun.*, 1997, 467.
[6]   Y. Tobe, S. Sasaki, M. Mizuno, K. Hirose and K. Naemura, *J. Org. Chem.*, 1998, **63**, 7481.

and X-ray structural analysis were all employed for structural characterisation. The association constant for dimerisation in the above solvent at 50 °C is 1.4×10^3 dm^3 mol^{-1}. In this study, a range of related thiourea derivatives were also demonstrated to exhibit broadly similar solution behaviour. The magnitude of the respective K_a values were found to depend on the steric bulk of the side chains, the acidity of the thiourea groups and, for some derivatives, the presence of weak intermolecular interactions between aromatic rings present in the side chains and the *m*-xylylene spacer group.

Overall, it should be noted that host–guest interactions are often characterised by unfavourable entropies of association (loss of disorder/degrees of freedom) – especially when the host and/or guest are not fully preorganised for binding to each other. For example, a receptor molecule composed of two 2-amino-6-methylpyridine groups connected by a terephthaloyl spacer has been demonstrated to form a complex with glutaric acid of structure **2** in 5% tetrahydrofuran in deutero-chloroform.[7] Under these conditions and at 295 K, the K_a is 6.4×10^2 dm^3 mol^{-1}. The

2

weak solvation of the hydrogen bonding sites of host and guest in the non-polar medium employed (indirectly) contributes to a strongly favourable enthalpic term for host–guest binding of $\Delta H = -33$ kJ mol^{-1}. However, the entropy term for the interaction is unfavourable (negative) ($\Delta S = -60$ J mol^{-1} K^{-1}), presumably chiefly reflecting the loss of translational and rotational motion accompanying bimolecular association, together with a contribution from the freezing of bond rotations on forming the complex. In accordance with the above, addition of dimethyl sulfoxide to a solution of **2** in a non-polar medium gives rise to strong solvation of the hydrogen bond donor sites, resulting in virtually complete dissociation of the complex.

In a subsequent study, the synthesis of urea and thiourea derivatives of types **3a** and **3b** were undertaken in an attempt to obtain new hosts that might yield more stable complexes with glutamic acid (of types **4a** and **4b**) and that might, in fact, 'hold together' in polar solvents.[8] The rationale for the design was based on the previously documented observation that both 1,3-dimethylurea and 1,3-dimethylthiourea form stable host–guest complexes with acetate in deutero-dimethyl sulfoxide; the K_a values for the latter are 45 dm^3 mol^{-1} and 340 dm^3 mol^{-1}, respectively.

7   F. Garcia-Tellado, S. Goswami, S.K. Chang, S. Geib and A.D. Hamilton, *J. Am. Chem. Soc.*, 1990, **112**, 7393.
8   E. Fan, S.A. Van Arman, S. Kincaid and A.D. Hamilton, *J. Am. Chem. Soc.*, 1993, **115**, 369.

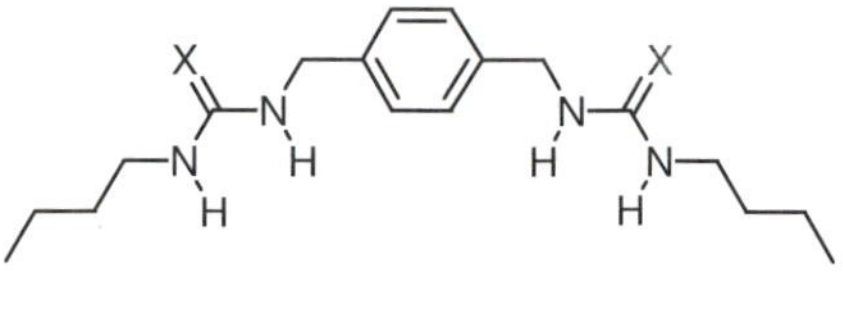

3a; X=O
3b; X=S

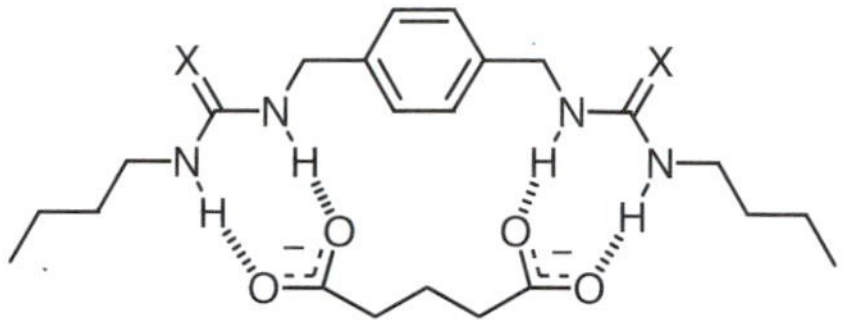

4a; X=O
4b; X=S

In contrast to **2**, the above derivatives incorporate four *hydrogen bond donors*.[9] As a consequence, a favourable bonding situation is present in which the four secondary amine protons of each host can interact with the two charged carboxylate groups of the guest – see **4a** and **4b**. The proposed structure of the complex of the bis-urea derivative was supported by the existence of large NMR downfield shifts for both the inner and outer 'urea' NH resonances in deutero-dimethyl sulfoxide and the observation of intramolecular ^{1}H NOEs between the receptor aryl and the guest CH$_2$ resonances in this solvent. A Job's plot confirmed the 1 : 1 stoichiometry of the product.

A further contribution to the stability of the complexes of types **4a** and **4b** was postulated to arise from the positioning of hydrogen bonding donor sites adjacent to each other in the receptor; for steric reasons these are less likely to be as effectively solvated relative to separated sites. Overall, the study illustrates well that the manipulation of both the location and charge of hydrogen bonding sites in synthetic receptors can result in a very large change in the binding strength towards a given guest.

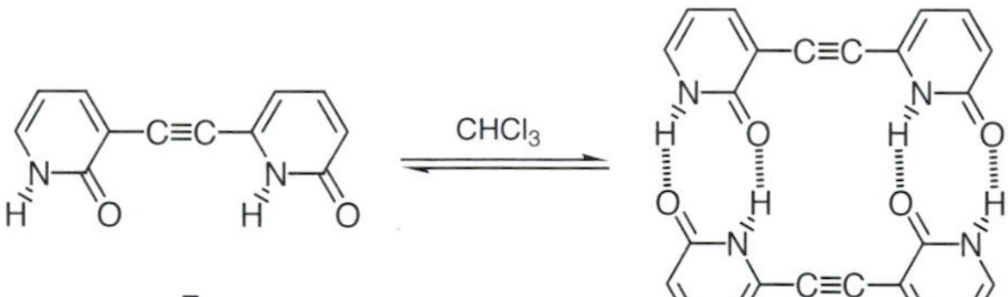

5

6

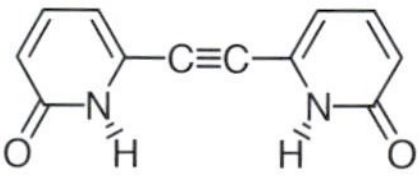

7

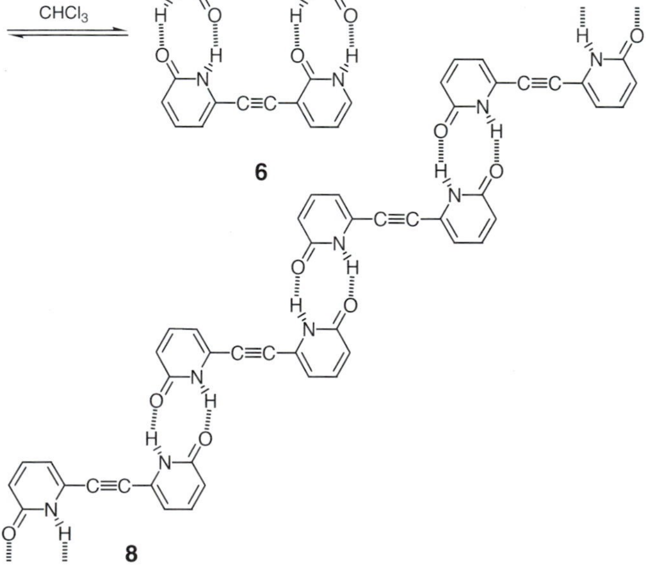

8

[9]   W.L. Jorgensen and J. Pranata, *J. Am. Chem. Soc.*, 1990, **112**, 2008; T.J. Murray and S.C. Zimmerman, *J. Am. Chem. Soc.*, 1992, **114**, 4010.

Structure **5** is one example of a number of dipyridones that incorporate different spacer groups.[10,11] Since **5** was designed to be self-complementary, it was anticipated that it would self-associate to produce a dimer of type **6**.[10] Indeed, this was shown to be so in chloroform (> 90% dimer) by means of vapour pressure osmometry. X-Ray crystallography also confirmed that the dimer persists in the solid state. The behaviour of **5** contrasts markedly with that of **7** which was designed to be complementary only in an 'offset' manner, such that linear polymerisation might be promoted. Under the dilute conditions of measurement, vapour pressure osmometry indicated that this species remains predominantly monomeric in chloroform; however, X-ray diffraction confirmed that **7** adopts the linear polymeric structure illustrated by **8** in the solid state. As anticipated, since self-association involves hydrogen bonding, both **5** and **7** were shown to exist only as monomers in the protic solvent methanol.

In an extension of the above study, the solution behaviour of the dipyridone **9**,

incorporating a flexible amine linker, was investigated.[11] Vapour pressure osmometry indicated that this species is predominantly dimeric in chloroform. This result was supplemented by NOE NMR experiments which suggested that dimerisation occurs to yield solution structures of type **10** and/or **11** under the conditions employed.

Hamilton *et al.*[12] have designed host molecules incorporating two linked 2,6-pyridinediamide groups which are beautifully aligned to complement the three carbonyl acceptor sites and two NH donor sites present in a barbiturate guest. Consequently, very stable complexes of type **12**, containing six intermolecular hydrogen bonds, were shown to form on mixing a host of this type and a barbiturate derivative as guest. As an extension of this study, the Hamilton group attempted to link redox or photoactive chromophores to individual hydrogen bonding host–guest units of the type just mentioned. In this manner, electronic or energy communication between the chromophores might be induced.

In a study of this type, the porphyrin-containing guest **13** was employed together with the dansyl derivative **14** as the host. The resulting complex of type **15** was

[10]   Y. Ducharme and J.D. Wuest, *J. Org. Chem.*, 1988, **53**, 5787.
[11]   M. Gallant, M.T.P. Viet and J.D. Wuest, *J. Org. Chem.*, 1991, **56**, 2284.
[12]   P. Tecilla, R.P. Dixon, G. Slobodkin, D.S. Alavi, D.H. Waldeck and A.D. Hamilton, *J. Am. Chem. Soc.*, 1990, **112**, 9408.

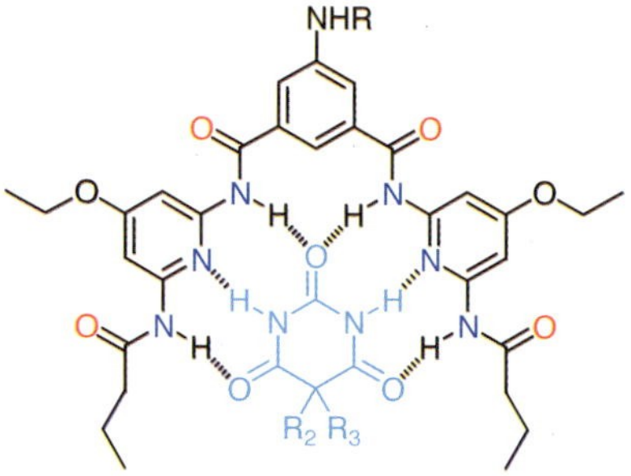

12

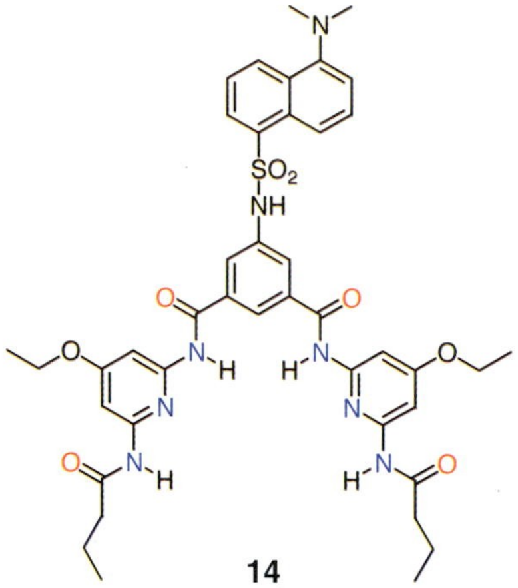

14

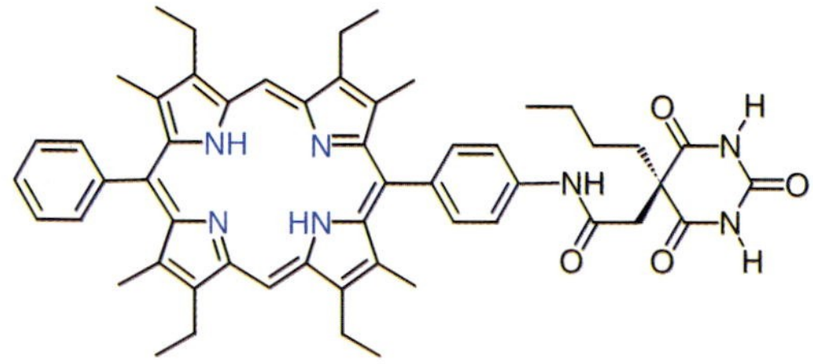

13

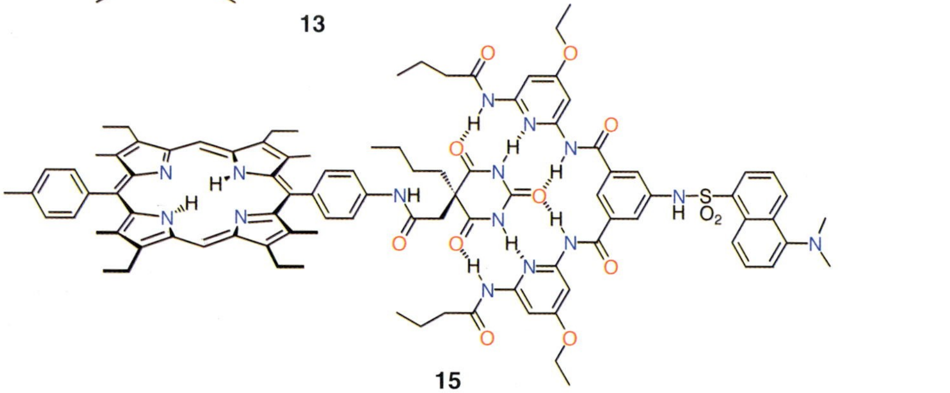

15

estimated to have the respective chromophores separated by about 23 Å. The degree of communication between these centres was assessed with the aid of fluorescence spectroscopy. Titration of the porphyrin-containing component into a dichloro-methane solution of **14** led to quenching of the dansyl fluorescence at 528 nm. A plot of emission intensely against concentration of **13** reveals a non-linear decrease in fluorescence as a function of complex concentration; this, together with the results from other (competition) experiments, was taken as evidence for occurrence of energy transfer quenching, implying significant communication between the chromophoric centres.

The dimerisation of macrocycles of type **16** in deutero-chloroform has been investigated using ¹H NMR and vapour pressure osmometry.[13] Macrocycles of this type

13   A.S. Shetty, J. Zhang and J.S. Moore, *J. Am. Chem. Soc.*, 1996, **118**, 1019.

R = COO(*n*-Bu)

16

are ideal for such studies since the large size and rigid nature of each ring results in the collective interactions between them being much stronger than for smaller flexible entities. They are thus more readily investigated. The π–π nature of the association in such cases is reflected by upfield NMR shifts of the aromatic proton signals, as expected from the presence of adjacent aromatic groups in the dimeric pair. Aromatic solvents have been well documented to interfere with π-stacking interactions because the solvent molecules are effective in solvating the solute. The absence of concentration-dependent chemical shift changes for (**16**; R = *n*-Bu) in deutero-benzene thus supports the predicted π-driven association in this case.

The study also demonstrated that apparently minor variations in the nature of R in particular cases may significantly alter the observed solution aggregation behaviour – confirming the sensitivity of the π-interactions to steric and electronic influences. Electron withdrawing substituents on the macrocycle tended to favour self-association when compared with the influence of electron donating substituents. Bulky (branched) groups positioned close to the aromatic core were found to inhibit association sterically. Finally, it was noted that stacking between triple bonds and aromatic rings, or between pairs of triple bonds, appear not to be particularly significant interactions in the above systems.

A good example of the manner by which a combination of hydrogen bonds and π–π interactions may provide an effective driving force for host–guest formation is given by the formation of the complex between *p*-benzoquinone and the tetraamide-containing cyclic host shown in Figure 3.1.[14] In this case the proposed host–guest interaction sites are illustrated in the insert to the figure. The polarisation of *p*-benzoquinone results in four edge-to-face π–π interactions on the quinone periphery being especially favourable. Indeed, the host is highly preorganised – the cavity appears perfectly complementary to the molecular structure of *p*-benzoquinone. NMR studies were in accord with the predicted structure of the adduct.

---

[14] C.A. Hunter, *J. Chem. Soc., Chem. Commun.*, 1991, 749.

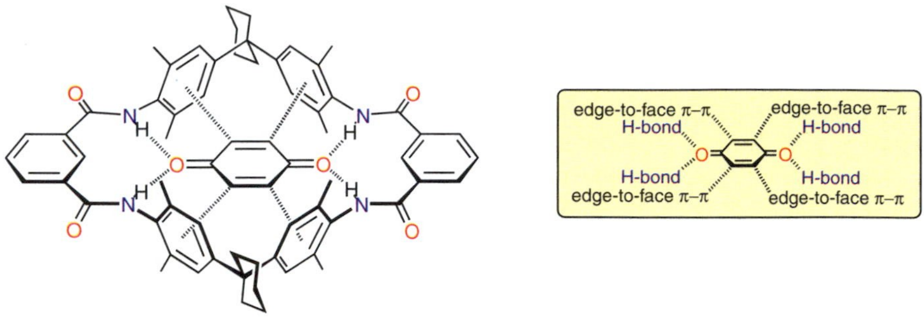

Figure 3.1 *The host–guest complex between the tetraamide macrocyclic host shown and benzoquinone*[14]

Titration of guest into host in deutero-chloroform also confirmed the 1 : 1 stoichiometry of the complex and yielded an association constant of 1.2×10^{-3} dm^3 mol^{-1}. Exchange between host and guest is fast on the ^{1}H NMR timescale. Not surprisingly, the macrocyclic host binds *p*-benzoquinone specifically; no complexation by this host was detected in the presence of a large excess of tetramethylbenzoquinone, tetrachlorobenzoquinone or anthraquinone.

3.2.2   Molecular Assembly as a Reaction Template

Examples of 1 : 2 host–guest systems capable of promoting a 'template' reaction between bound guests have been reported.[15] In these, the receptor incorporates two separate hydrogen bonding regions. The presence of the latter promote the self-assembly of a tertiary complex in which the guests are aligned such that intermolecular condensation is facilitated. A system of this type is illustrated in Figure 3.2. In this case, for simplicity, the mechanistically straightforward S$_N$2 alkylation of an amine by an alkyl halide was selected for study. The template in this case possesses non-identical binding sites. Control experiments were in accord with the formation of the ternary complex as an intermediate; a relatively modest reaction rate enhancement of 12-fold was observed.

3.2.3   Larger Linked Systems

Di- and polytopic host–guest systems have provided a convenient starting point for the construction of larger assemblies and many systems of this type are now known. For example, in an early study Kimura *et al.*[16] synthesised the catecholamine complex **17**. The crown ether unit of this ditopic host was known to be an effective receptor for primary alkyl ammonium salts, whereas the partially protonated form of the hexamine ring had been documented to bind anionic substrates (such as carboxylates) or electron-donor substrates (such as catechols). Accordingly, this host forms stable 1 : 1 complexes with zwitterionic guests such as amino acids, pep-

[15]   T.R. Kelly, G.J. Bridger and C. Zhao, *J. Am. Chem. Soc.*, 1990, **112**, 8024.
[16]   E. Kimura, H. Fujioka and M. Kodama, *J. Chem. Soc., Chem. Commun.*, 1986, 1158.

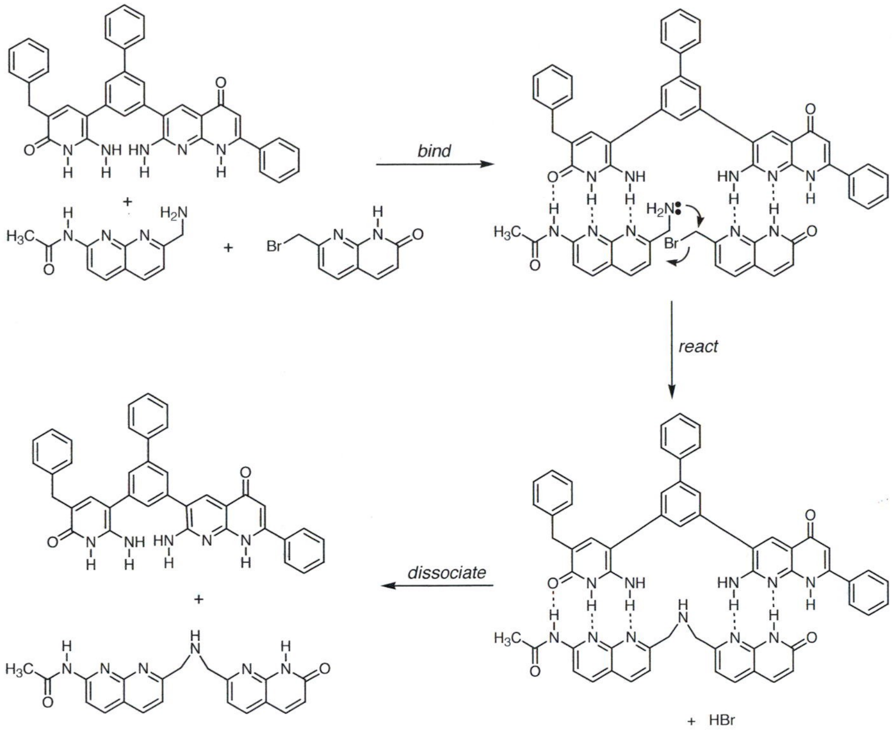

Figure 3.2 *Molecular assembly used as reaction template to promote intermolecular condensation of two bound guests*[15]

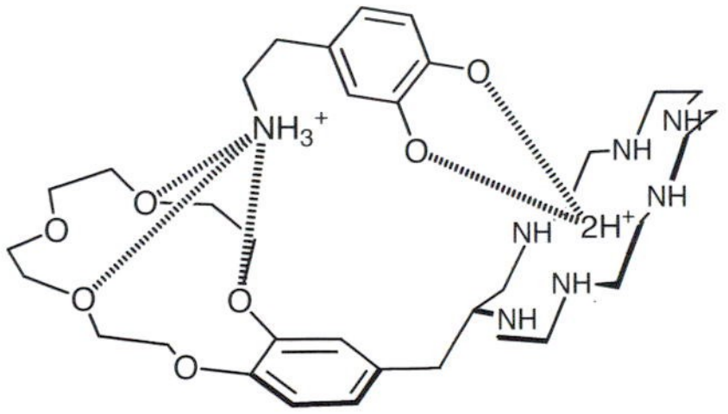

17

tides and catecholamine (as shown) in aqueous solution at neutral pH (where the macrocyclic amine groups are partially protonated).

Whitesides *et al.* have prepared a series of large self-assembled structures containing multiple host–guest interactions.[17] These assemblies were built around primary interactions between cyanuric acid (CA) and melamine (M) derivatives as illustrated by **18** (and were designated CA·M); where CA contains similar hydrogen

[17] C.T. Seto and G.M. Whitesides, *J. Am. Chem. Soc.*, 1990, **112**, 6409.

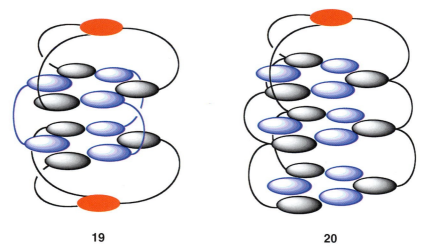

18

bonding sites to those of barbiturate). It had been proposed previously that a highly symmetrical, infinite two-dimensional lattice of type $(CA \cdot M)_n$ forms spontaneously on mixing the respective components.[17]

The structure of the first product in this series of assemblies is illustrated in Figure 3.3.[17] In this case a central 'hub' with three radiating flexible 'spokes' was employed to covalently link three melamine derivatives. The resulting tritopic receptor ('rosette') is abbreviated $hubM_3$. This species binds three neohexylcyanurate guests such that the overall stoichiometry of the host–guest assembly is 1 : 3 $(hubM_3 \cdot CA_3)$.

Further members of the series having, for example, double-decker (**19**) and triple-decker (**20**) structures have been prepared together with other examples incorpo-

19	**20**

rating three separate components.[18] Structures incorporating both linked melamines and linked isocyanurates have also been characterised. For example, the species $hub(MM)_3$, containing six melamines preorganised to recognise six isocyanurates (CAs), reacts with monomeric, dimeric and trimeric derivatives of isocyanuric acid to yield self-assembled aggregates displaying various stoichiometries.[19] These

18   C.T. Seto and G.M. Whitesides, *J. Am. Chem. Soc.,* 1991, **113**, 712; C.T. Seto, J.P. Mathias and G.M. Whitesides, *J. Am. Chem. Soc.,* 1993, **115**, 1321; C.T. Seto and G.M. Whitesides, *J. Am. Chem. Soc.,* 1993, **115**, 1330; D.N. Chin, D.M. Gordon and G.M. Whitesides, *J. Am. Chem. Soc.,* 1994, **116**, 12033; G.M. Whitesides, E.E. Simanek, J.P. Mathias, C.T. Seto, D.N. Chin, M. Mammen and D.M. Gordon, *Acc. Chem. Res.,* 1995, **28**, 37.

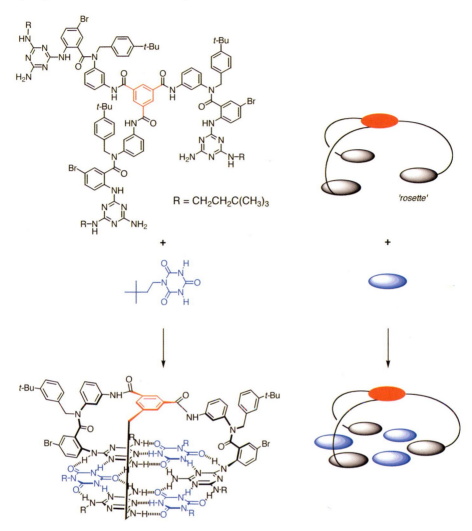

Figure 3.3 *Formation of a tritopic receptor in 'rosette' form by self-assembly*[17]

soluble products comprise between four and seven individual molecules and the operation of positive co-operativity in the self-assembly process has been demonstrated in one case.

A related aggregate based on 'hub(MMM)$_3$', incorporating nine melamine centres, has been reported.[20] In this case, hub(MMM)$_3$ associates with nine neohexylisocyanurates to yield an aggregate containing ten molecules that are arranged in three parallel rosettes. The structure is stabilised by 54 hydrogen bonds.

19   J.P. Mathias, C.T. Seto, E.E. Simanek and G.M. Whitesides, *J. Am. Chem. Soc.*, 1994, **116**, 1725.
20   J.P. Mathias, E.E. Simanek, C.T. Seto and G.M. Whitesides, *Angew. Chem., Int. Ed. Engl.,* 1993, **32**, 1766.

NMR spectroscopy (including NMR titrations), gel permeation chromatography and vapour pressure osmometry have proved a useful combination for establishing the structures and stabilities of the respective aggregates. Clearly, particular members of the series are stabilised by the additivity of multiple hydrogen bonds as well as by a reduction in the entropies of translation that would normally oppose self-assembly when single (non-linked) components assemble in this manner. For example, the enthalpy associated with the formation of 36 hydrogen bonds in two parallel connected 'rosettes' in the single supramolecular complex of type $hub(MM)_3 \cdot (neohexCA)_6$ is clearly sufficient to over-ride the unfavourable entropy arising from the association of the seven entities from which it is composed. In accord with this, aggregates based on two connected 'rosettes' were shown to be more stable than related species based on a single 'rosette'.[21]

It is also clear from these studies that steric hindrance between substituents in the hydrogen-bonded networks is sufficient to 'programme' the selection of the cyclic 'rosette' motif in preference to competing linear, hydrogen bonding arrangements.[21]

3.2.4   Molecular Tweezers

Besides those discussed already, other systems containing molecular clefts – designed for guest binding – have been reported. These include several examples of molecular 'tweezers'.[22] An example prepared by successive Diels–Alder reactions is represented by **21**.[23] Because of its overall dimensions and its semi-rigid, ribbon-like structure, it was anticipated that **21** should be appropriately preorganised for complexation of aromatic guest molecules through multiple arene–arene interactions. Further, the possibility of bond angle distortions should permit a cer-

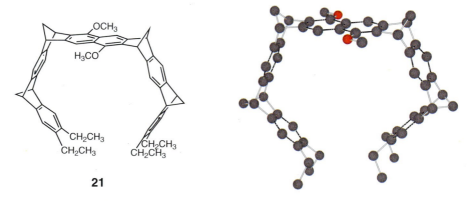

21

MM2 minimised structure of **21**

[21]   J.P. Mathias, E.E. Simanek and G.M. Whitesides, *J. Am. Chem. Soc.*, 1994, **116**, 4326.
[22]   See, for example: C.W. Chen and H.W. Whitlock, *J. Am. Chem. Soc.*, 1978, **100**, 4921; M. Harmata, C.L. Barnes, S.R. Karra and S. Elahmad, *J. Am. Chem. Soc.*, 1994, **116**, 8392; C. Valdés, U.P. Spitz, L.M. Toledo, S.W. Kubik and J. Rebek, *J. Am. Chem. Soc.*, 1995, **117**, 12733; M. Kamieth and F.-G. Klärner, *J. Prakt. Chem.*, 1999, **341**, 245.
[23]   F.-G. Klärner, J. Benkhoff, R. Boese, U. Burkert, M. Kamieth and U. Naatz, *Angew. Chem., Int. Ed. Engl.*, 1996, **35**, 1130.

tain degree of flexibility in this system, allowing the tweezer arms to expand or contract such that interaction with a suitable guest can be optimised. ^{1}H NMR experiments in deutero-chloroform indicated that **21** forms complexes with 1,4-dicyanobenzene or 1,4-benzoquinone. In each case an upfield shift of the guest's aromatic signal was observed, indicating that the hydrogen atoms are located in the magnetic anisotropic regions of the host–arene units.

Interestingly, when **21** was recrystallised from a 2 : 1 mixture of cyclohexane and ethyl acetate, a 1 : 1 complex formed containing a disordered cyclohexane molecule. The orientation of this guest seemed to correlate with the presence of disorder in peripheral substituents on the host. It was concluded that even the relatively weak, but still apparently somewhat directed, alkane–arene interactions are of sufficient strength to promote formation of a host–guest complex in this case.

3.3 Cyclic Assemblies

Covalently-linked cyclophanes have long provided the basis for a rich source of inclusion chemistry involving a wide variety of guest types.[24] Increasingly, self-assembly has been employed to produce cyclophane-like entities that are held together by non-covalent interactions. A theoretical treatment of such a self-assembly macrocyclisation process starting from a difunctional monomer, and occurring under thermodynamic control, has been presented.[25] The parameters on which this treatment is based are the effective molarity (EM_n) of the self-assembling cyclic *n*-mer and the equilibrium constant (K_{inter}) for the intermolecular model reaction involving the corresponding mono-functional reagents. Using these quantities, the distribution curve for formation of the self-assembled macrocycle has been modelled.

It was predicted that, in the limit of an infinite value of K_{inter}, there is a critical monomer concentration ($cmc = nEM_n$) below which the solution is virtually solely composed of the self-assembled macrocycle and above which the concentration of the latter remains constant, with the excess monomer then producing only non-cyclic species.

Considerable interest has been shown in the formation of hydrogen bond-mediated (discrete) cyclic assemblies incorporating between three and ten component molecules.[26] A few hydrogen bonding motifs have dominated these studies.[27] These include: cyanuric acid–melamine contacts, 2-aminopyridine–carboxylic acid contacts and carboxylic acid or pyridone dimer contacts. While individual hydrogen bonding interactions of this type are often weak (with association constants of

[24] F.N. Diederich, *Cyclophanes*, Royal Society of Chemistry, Cambridge, 1991; F.N. Diederich, in *Supramolecular Chemistry*, V. Balzani and L. De Cola, eds., Kluwer, Dordrecht, 1992, pp. 119–136; F.V. Vögtle, *Cyclophane Chemistry*, John Wiley and Sons, Chichester, 1993.
[25] G. Ercolani, *J. Phys. Chem. B*, 1998, **102**, 5699.
[26] S.V. Kolotuchin and S.C. Zimmerman, *J. Am. Chem. Soc.*, 1998, **120**, 9092.
[27] See, for example: R.H. Vreekamp, J.P.M. van Duynhoven, M. Hubert, W. Verboom and D.N. Reinhoudt, *Angew. Chem., Int. Ed. Engl.*, 1996, **35**, 1215; S.C. Zimmerman, F. Zeng, D.E.C. Reichert and S.V. Kolotuchin, *Science*, 1996, **271**, 1095; S.C. Zimmerman and B.F. Duerr, *J. Org. Chem.*, 1992, **57**, 2215; E. Boucher, M. Simard and J.D. Wuest, *J. Org. Chem.*, 1995, **60**, 1408; J. Yang, E.K. Fan, S.J. Geib and A.D. Hamilton, *J. Am. Chem. Soc.*, 1993, **115**, 5314.

around 10^2 dm^3 mol^{-1} in deutero-chloroform), at appropriate concentrations the cyclic assemblies are nevertheless expected to be favoured enthalpically relative to corresponding non-cyclic ones.

3.3.1 Some Cyclic Oligomers

It has been demonstrated that simple 5-substituted isophthalic acid derivatives form well-defined cyclic hexameric aggregates of type **22** in the solid state that are

22

stabilised by 12 hydrogen bonds.[28] Vapour pressure osmometry strongly suggested that such hexameric units are preserved in toluene solution at concentrations above 10^{-2} mol dm^{-3}.

The very stable cyclic hexamer **23**, assembled from the corresponding monomer incorporating donor–donor–acceptor and acceptor–acceptor–donor motifs, was demonstrated to form spontaneously in deuterated tetrahydrofuran, chloroform and toluene.[26] Molecular modelling suggested that the methyl groups pointing to the centre of the structure alternate up and down out of the main plane of the assembly such that S_6 symmetry is achieved and all methylene protons are diastereotopic. A feature of the self-complementary hydrogen bonding sites as arranged in the cyclic structure is the prospect that six (secondary) hydrogen bonds may also form; one of the NH groups in each monomer appears able to serve as a long-range donor to an opposing heterocyclic N in the next, as shown in **23**. Molecular weight determinations were performed using size exclusion chromatography – the results were in accord with the proposed structure. ^{1}H NMR experiments using a range of

[28]   J. Yang, J.-L. Marendaz, S.J. Geib and A.D. Hamilton, *Tetrahedron Lett.*, 1994, **35**, 3665.

23

(aprotic) solvents, concentrations and temperatures confirmed the inherent stability of this strongly hydrogen-bound ring system.

In a related manner, other hydrogen bonding motifs have been employed to produce cyclic hexamers. Structure **24** provides a further example of this type.[29] This product was formed in warm dimethyl sulfoxide from the corresponding monomer. The latter incorporates the acceptor–acceptor–donor sequence of cytosine and the complementary donor–donor–acceptor sequence of guanine mutually orientated at 120°. In this case **24** was isolated as large, colourless crystals and X-ray diffraction (employing synchrotron radiation) confirmed its hexagonal arrangement (Figure 3.4).

3.3.2 Molecular Boxes

Molecular 'boxes', capable of binding a decanediyldiammonium or dodecanediyldiammonium ion in their cavities, have been constructed from diazacrown ethers and pendant nucleotide bases.[30] For example, structure **25** is based on hydrogen-bonded adenine–diazacrown–adenine and thymine–diazacrown–thiamine monomer units. The bis(ammonium) salt interacts with the oxygen atoms of two azacrown rings contained in each half of the cyclic 1 : 1 host–guest 'box'. Characterisation of this product was largely performed using ¹H-NMR spectroscopy and vapour

29 M. Mascal, N.M. Hext, R. Warmuth, M.H. Moore and J.P. Turkenburg, *Angew. Chem., Int. Ed. Engl.*, 1996, **35**, 2204.
30 M.S. Kim and G.W. Gokel, *J. Chem. Soc., Chem. Commun.*, 1987, 1686; O.F. Schall and G.W. Gokel, *J. Am. Chem. Soc.*, 1994, **116**, 6089.

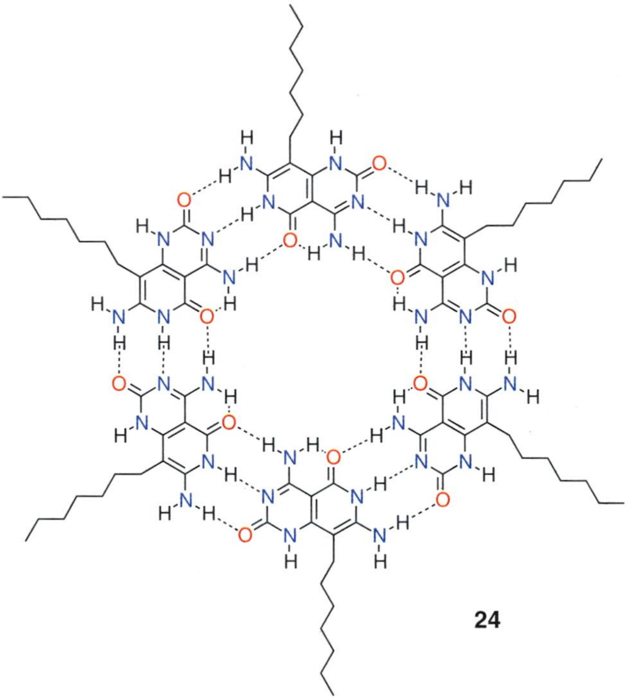

24

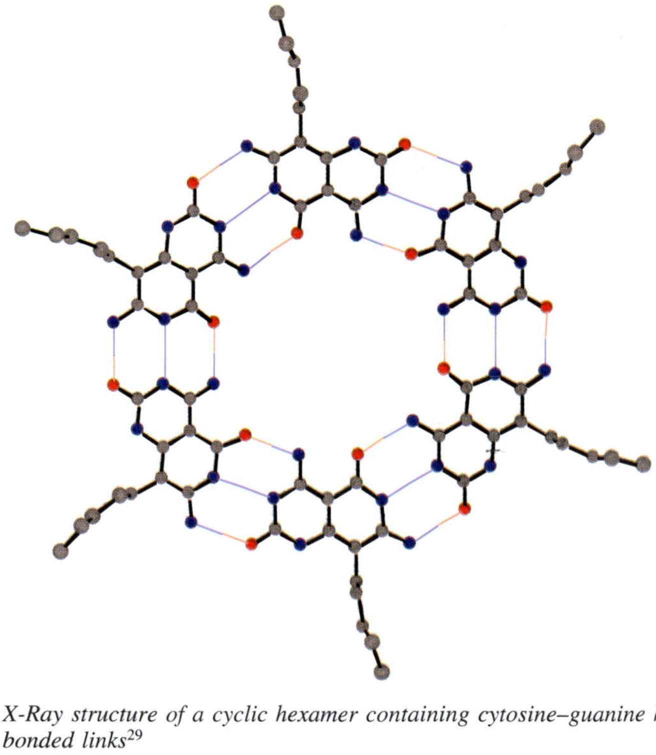

Figure 3.4 *X-Ray structure of a cyclic hexamer containing cytosine–guanine hydrogen bonded links*[29]

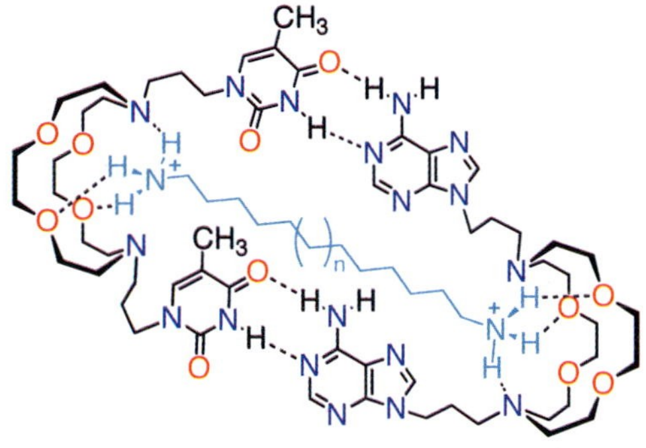

25

pressure osmometry, and it was demonstrated that Hoogsteen hydrogen bonding plays a significant role in both inter- and intra-molecular interactions used to form the dimeric box structure. The 1 : 1 association constant between the complementary diazacrown-containing halves of the cyclic dimer was found to be 855 dm³ mol⁻¹ (deutero-chloroform). In the presence of the diammonium salt, the box appears to assemble around it, ultimately acting as a ditopic receptor.

The bis-porphyrin supramolecular box **26** has also been assembled using a molecular recognition process involving hydrogen bonding between pendant-arm

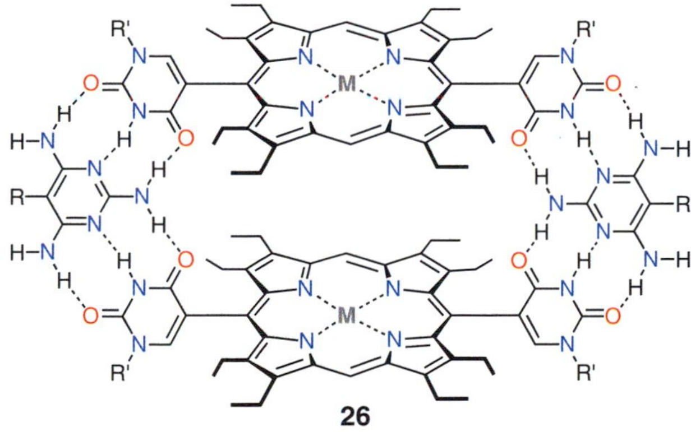

M = 2H⁺, Fe(III), Zn(II)
R = octyl or decyl
R' = decyl

26

uracil sites on the porphyrin rings and complementary 'free' alkyl-substituted tri-aminopyridinepyrimidines which act as bridges.[31] The uracil groups are linked to each porphyrin such that rotation about the connecting bond is hindered by the ethyl substituents on the porphyrin pyrroles. Thus, the substituted porphyrin exists in two rotameric forms, *syn* and *anti*, with respect to the relative orientation of the two uracil groups. Evidence for self-assembly was obtained from a variety of measurements including NMR, electrospray mass spectrometry, vapour pressure osmometry as well as fluorescence quenching and enhancement. Addition of the alkyl

[31]   C.M. Drain, R. Fischer, E.G. Nolen and J.-M. Lehn, *J. Chem. Soc., Chem. Commun.*, 1993, 243.

triaminopyrimidine led to a five- to seven-fold increase in the solubility of the porphyrin species in aprotic solvents. The existence of two rotamers in the case of the
latter species was confirmed by the presence of signal doubling in the ¹H NMR
spectrum. On assembling the cage in dichloromethane, there is evidence that the
syn/anti equilibrium is slowly shifted towards the *syn* form. Nevertheless, structure
26 assembles directly from the *syn* isomer, whereas the *anti* isomer probably generates a zig-zag structure. Interestingly, the di-zinc form of **26** binds a 4,4′-bipyri-
dine ligand more strongly than comparable, substituted zinc porphyrin monomeric
units. Structure **26** represents a self-assembled, hydrogen-bonded analogue of a
number of covalently-linked cofacial porphyrin dimers reported previously.[32]

Three equivalents of a calix[4]arene derivative, diametrically substituted at its
upper rim with two melamine units, have also been demonstrated to yield well-
defined box-like assemblies on interaction with six equivalents of 5,5-diethylbar-
bituric acid.[33] The product, which is based on an extended 'rosette' motif, consists
of nine different components held together by 36 hydrogen bonds. It is stable in
apolar solvents and remains so even at very low concentrations. Further, the structure remains intact in the solid state, as confirmed by an X-ray diffraction study.
In an extension of this study, three enantio-pure calix[4]arene–dimelamine derivatives were employed for a parallel study.[34] In this instance, the chiral information
in the individual calixarene derivatives proved successful in controlling the
conformation of the final assembly – a homochiral supramolecular entity was the
result.

3.4   Cylindrical Assemblies

There has been a number of examples of cylindrical (tubular) supramolecular structures reported over recent years. These encompass all-carbon structures as well as
other inorganic and organic tubes (and columns) and include lipid-based systems.[35]
However, the majority of these tubes represent extended arrays (on a multi-nanome-
tre or higher scale) and hence fall outside the scope of the present discussion – in
any case, many were not obtained directly by self-assembly processes. Indeed, only
a limited number of self-assembled, discrete cylindrical systems have been
described.

Bonar-Law and Saunders[36] have demonstrated that 'cyclocholates' (see **27**; Figure
3.5) reversibly self-assemble in organic solvents to form short, molecular cylinders

[32]   See, for example: A. Hamilton, J.-M. Lehn and J.L. Sessler, *J. Am. Chem. Soc.*, 1986, **108**, 5158;
       A. Osuka, F. Kobayashi and K. Maruyama, *Bull. Chem. Soc., Jpn.*, 1991, **64**, 1213 and references
       therein.
[33]   P. Timmerman, R.H. Vreekamp, R. Hulst, W. Verboom, D.N. Reinhoudt, K. Rissanen, K.A.
       Udachin and J. Ripmeester, *Chem. Eur. J.*, 1997, **3**, 1823.
[34]   L.J. Prins, J. Huskens, F. de Jong, P. Timmerman and D.N. Reinhoudt, *Nature*, 1999, **398**, 498.
[35]   See, for example: N. Khazanovich, J.R. Granja, D.E. McRee, R.A. Milligan and M.R. Ghadiri, *J.
       Am. Chem. Soc.*, 1994, **116**, 6011 and references therein; A. Ikeda and S. Shinkai, *J. Chem. Soc.,
       Chem. Commun.*, 1994, 2375; P.M. Ajayan, O. Stephan, Ph. Redlich and C. Colliex, *Nature*, 1995,
       375, 564; M. Nishizawa, V.P. Menon and C.R. Martin, *Science*, 1995, **268**, 700; V. Percec, D.
       Schlueter, G. Ungar, S.Z.D. Cheng and A. Zhang, *Macromolecules*, 1998, **31**, 1745.
[36]   R.P. Bonar-Law and J.K.M. Sanders, *Tetrahedron Lett.*, 1993, **34**, 1677.

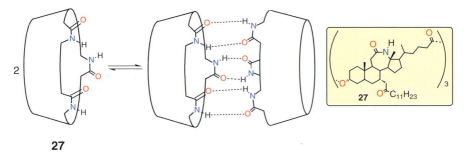

27

Figure 3.5 *Reversible self-assembly of a short molecular cylinder through dimerisation of the cyclic chloric derivative* **27**[36]

of the type shown in the figure. Assembly (dimerisation) occurs through hydrogen bond formation involving the complementary *cis*-amide functions on the rim of each macrocycle. Equilibrium constants were obtained by both ¹H NMR and/or FT-IR experiments in dry carbon tetrachloride. The NMR method, in which the position of the fast-exchange (averaged) NH resonance was plotted against macrocycle concentration, yielded the more accurate results. A value of 3×10^4 dm³ mol⁻¹ was obtained for the dimerisation of **27**. Molecular weight measurement using vapour pressure osmometry yielded a value that corresponded to the presence of the dimer. Molecular mechanics calculations supported other evidence that the face-to-face hydrogen-bonded dimer is in fact the arrangement adopted in this case.

3.5 Spherical Assemblies

3.5.1 Self-assembled Cages

A variety of covalently-bound bicyclic molecules, capable of complexing and 'including' smaller guest species, have been reported over many years.[37]

More recently, a number of groups have employed strategies for achieving a similar result through self-assembly.[38] Such a prospect is attractive since the association step will often be reversible under relatively mild conditions. Rebek *et al.*[39,40] were successful in assembling smaller subunits to form new hydrogen-bonded, cavity-containing structures. A range of guests has been encapsulated in the resulting cages. In their initial study, the above group demonstrated the self-association of **28** to yield a dimeric assembly with a pseudo-spherical shape.[41,42] The internal

[37] L.F. Lindoy, *The Chemistry of Macrocyclic Ligand Complexes*, Cambridge University Press, Cambridge, UK, 1989; G.W. Gokel, *Crown Ethers and Cryptands*, Royal Society of Chemistry, Cambridge, UK, 1991; R. Berscheid, I. Lüer, C. Seel and F. Vögtle in *Supramolecular Chemistry*, V. Balzani and L. De Cola, eds., Kluwer, Dordrecht, 1992, pp. 71–86.
[38] J. de Mendoza, *Chem. Eur. J.*, 1998, **4**, 1373.
[39] J. Rebek, *Pure Appl. Chem.*, 1996, **68**, 1261.
[40] R.M. Grotzfeld, N. Branda and J. Rebek, *Science*, 1996, **271**, 487.
[41] R. Wyler, J. de Mendoza and J. Rebek, *Angew. Chem., Int. Ed. Engl.,* 1993, **32**, 1699.
[42] N. Branda, R. Wyler and J. Rebek, *Science*, 1994, **263**, 1267.

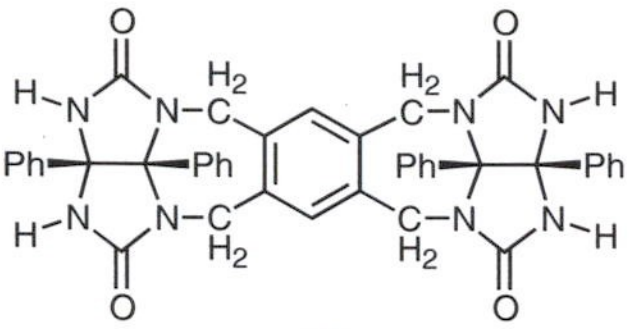

28

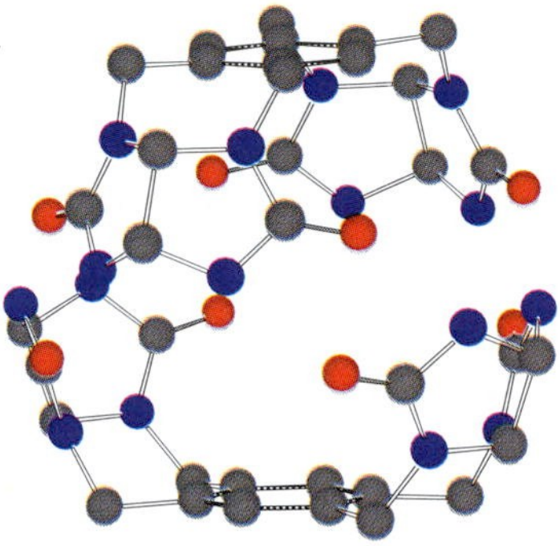

Phenyl groups omitted for clarity, MM2 minimised structure

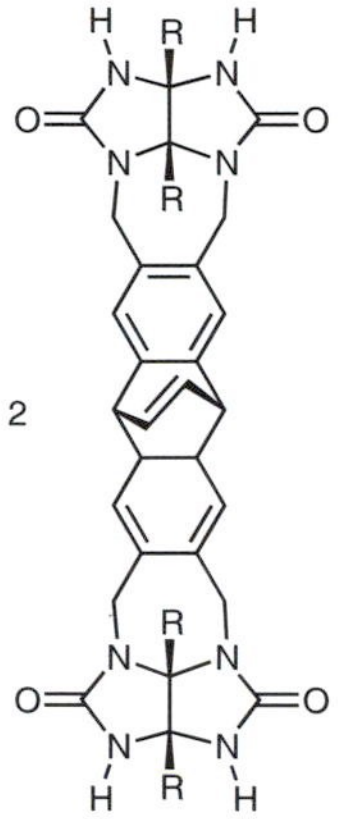

29

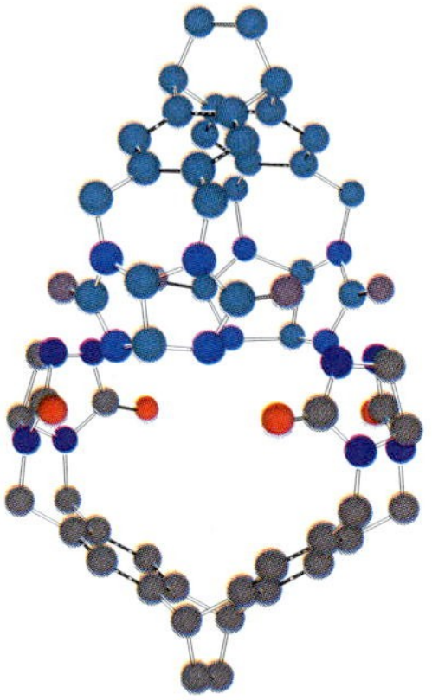

Glycouril ester groups omitted for clarity, MM2 minimised structure

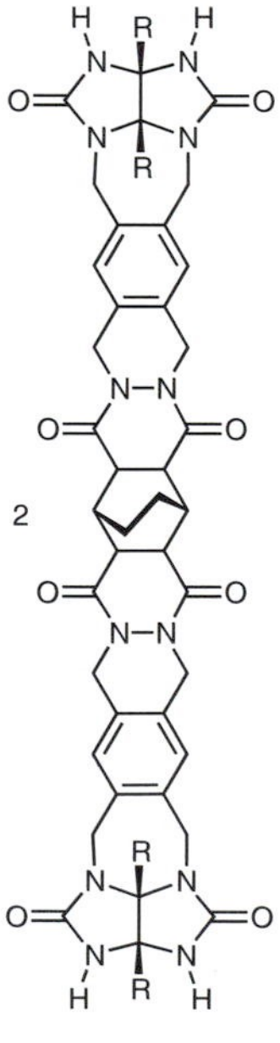

R = Phenyl
30

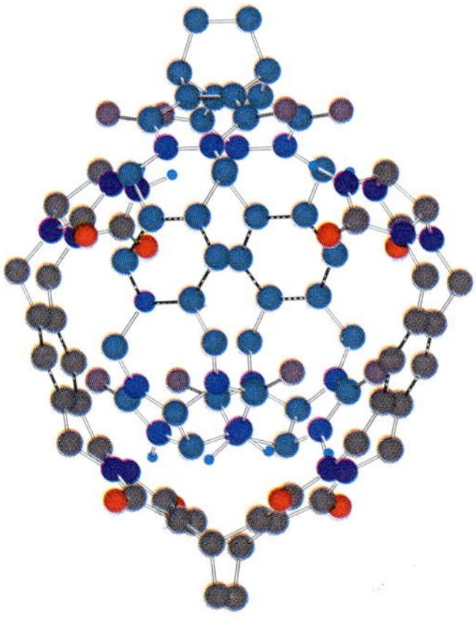

R groups are omitted for clarity, MM2 minimised structure

cavity of the product was found to act as a host for small guests provided they were of complementary shape and volume. The dimer is held together by eight hydrogen bonds aligned along a seam that resembles that of a tennis ball. In a weakly-competitive solvent such as deutero-chloroform, guests such as methane, ethane and ethylene have been encapsulated within the cavity.

Based on the above study, larger pseudo-spherical host assemblies have been constructed from extended tetra-urea derivatives; structures **29**[43] and **30**[44,45] are typical. Once again, each of these have been found to encapsulate suitable guests in solution. As before, guests that best complement the cavity in both size and electronic nature are the ones found to be preferentially encapsulated.[46,47] Consider **30**: it is crescent-shaped and incorporates 13 fused rings; it dimerises spontaneously in benzene to yield the corresponding spherical cage. Dimerisation results in the individual molecules of **30** being arranged mutually at right angles with their concave sides facing each other. In this arrangement the urea-derived end of one molecule is able to hydrogen bond with the carbonyl oxygens of the other. Guest solubility can be markedly affected: normally insoluble adamantane dicarboxylate is rapidly solubilised on encapsulation by this cage.

The generation of systems in which symmetrical molecules assemble through hydrogen bonding to produce capsules with dissymmetric cavities have also been reported.[48] Such capsules assemble and dissipate on a timescale that allows their direct observation using NMR measurements. The recognition of chiral guests, such as naturally-occurring terpenes, directs which dissymmetric cavities are preferentially formed in the assembly process.

From the same laboratory, other self-assembled dimeric systems, based on monomers incorporating glycoluril subunits, held apart by rigid backbone spacers, have been demonstrated to form. The resulting pseudo-spherical capsules are again held together by hydrogen bonds.[49] A monomer from this series is illustrated by **31** in Figure 3.6. The figure also illustrates the predicted structure **32** of the corresponding capsule. Related systems incorporating rigid ethylene, naphthalene and ethenoanthracene spacer groups comprise other members of the series. NMR studies revealed that all of these systems readily enclose suitable smaller guest molecules in a reversible manner. For example, dichloromethane was detected inside **32** using ^{13}C NMR while the related capsule incorporating an ethylene bridge in each component of the dimer was demonstrated to bind methane selectively in the presence of ethane.

'Hybrid' capsules composed of two glycoluril derivatives incorporating different backbone spacer groups have also been demonstrated to form in solutions containing a mixture of two 'homo-dimers' chosen from the above series. In this

43 C. Valdés, U.P. Spitz, S.W. Kubik and J. Rebek, *Angew. Chem., Int. Ed. Engl.*, 1995, **34**, 1885.
44 R.S. Meissner, J. Rebek and J. de Mendoza, *Science*, 1995, **270**, 1485.
45 J.M. Rivera, T. Martin and J. Rebek, *J. Am. Chem. Soc.*, 1998, **120**, 819.
46 Y. Tokunaga, D.M. Rudkevich and J. Rebek, *Angew. Chem., Int. Ed. Engl.*, 1997, **36**, 2656.
47 S. Mecozzi and J. Rebek, *Chem. Eur. J.*, 1998, **4**, 1016.
48 J.M. Rivera, T. Martin and J. Rebek, *Science*, 1998, **279**, 1021.
49 C. Valdés, U.P. Spitz, L.M. Toledo, S.W. Kubik and J. Rebek, *J. Am. Chem. Soc.*, 1995, **117**, 12733.

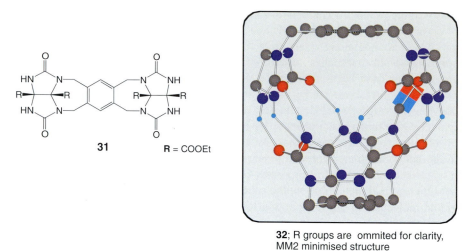

32; R groups are ommited for clarity,
MM2 minimised structure

Figure 3.6 *A pseudo-spherical capsule composed of bridged glycoluril subunits linked by hydrogen bonds*[49]

manner, capsules featuring cavities that offer a range of differing sizes and shapes were obtained.

Several extensions of the above study have now been reported leading to the synthesis of other pseudo-spherical 'softballs' of the above type. For example, two (identical) self-complementarity subunits of type **33**, also assemble to form a large,

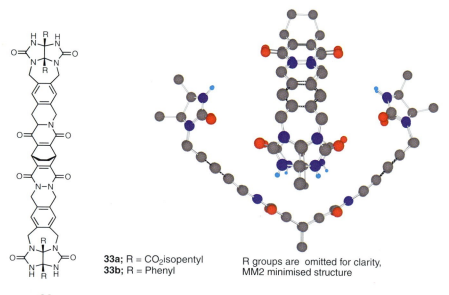

33a; R = CO₂isopentyl
33b; R = Phenyl

R groups are ommited for clarity,
MM2 minimised structure

33

hollow sphere.[50] Once again, the reversible encapsulation of smaller molecules, including two solvent benzene molecules, has been reported. In this latter case the included two solvent molecules may in turn be replaced by a single adamantane guest. The latter results in an increase in the number of 'free' molecules – reflected by a positive entropy term for the exchange process. In fact, in this case both the enthalpy and entropy terms increase, resulting in the overall process being entropy driven.

As also discussed in subsequent chapters, the torus-shaped cyclodextrins are naturally-occurring, cyclic oligosaccharides composed of six (α), seven (β), and eight (γ) linked α-1,4-linked D-glucopyranose subunits.[51] These unusual, water-soluble molecules typically form 1 : 1 complexes with hydrophobic guests of suitable dimensions. Although less common, examples of cyclodextrins forming 2 : 1 (cyclodextrin : guest) complexes are also known (see below).[52,53] Related to this, examples in which two or more cyclodextrin units have been linked covalently have been reported and there is evidence that such linking results in higher binding affinities when bound to a suitable hydrophobic guest.[51,54]

The formation of hydrophilic molecular capsules by self-assembly of two cyclodextrin entities in the presence of a guest has been investigated.[54] A system of this type was demonstrated to self-assemble from per-6-amino β-cyclodextrin (**34**) and per-6-thioglycolic β-cyclodextrin (**35**) in 50 mmol dm^{-3} KCl solution at neutral pH.[55] The structure of the resulting dimer has been investigated using a

34   **35**

dynamic ^{1}H NMR analysis experiment which, despite treating molecular species as smooth rigid bodies, was shown to provide useful information about the average structure (shape and volume) of the associated species. The most probable equilibrium distance between the interacting units in the dimer, as deduced from the NMR studies, is around 6 nm.

50 R. Meissner, X. Garcias, S. Mecozzi and J. Rebek, *J. Am. Chem. Soc.*, 1997, **119**, 77.
51 C.J. Easton and S.F. Lincoln, *Modified Cyclodextrins: Scaffolds and Templates for Supramolecular Chemistry*, Imperial College Press, London, 1999.
52 M.M. Conn and J. Rebek, *Chem. Rev.*, 1997, **97**, 1647.
53 R.G. Chapman and J.C. Sherman, *Tetrahedron*, 1997, **53**, 15911.
54 A. Jasat and J.C. Sherman, *Chem. Rev.*, 1999, **99**, 931.
55 B. Hamelin, L. Jullien, F. Guillo, J.-M. Lehn, A. Jardy, L. De Robertis and H. Driguez, *J. Phys. Chem.*, 1995, **99**, 17877; B. Hamelin, L. Jullien, C. Derouet, C.H. du Penhoat and P. Berthault, *J. Am. Chem. Soc.*, 1998, **120**, 8438.

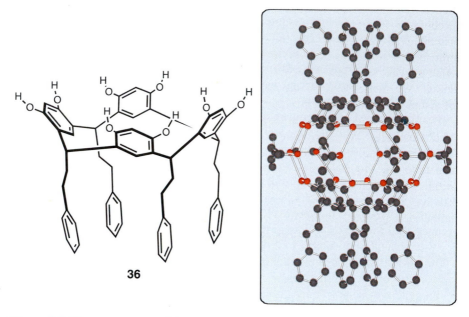

36

Figure 3.7 *The arrangement of the two resorcinarene monomers of type* **36** *in the corresponding dimeric capsule. The monomers are held together by a regular array of hydrogen bonds involving the phenolic hydrogens and eight propan-2-ol solvent molecules*[62]

Self-assembly and molecular encapsulation processes involving calix[4]arene derivatives have been the subject of a considerable number of studies.[52,56–62] Thus, large, dimeric capsules based on calix[4]arenes and the corresponding resorcinarenes, showing related behaviour to the systems discussed above, have been investigated. An example is given by the formation of the large carcerand-like dimer of calix[4]resorcinarene (**36**; Figure 3.7).[62] This species exists in the solid state and is held together in a 'rim-to-rim' fashion by an array of eight hydrogen-bonded propan-2-ol solvent molecules.[62] Single crystals were grown by slow fusion of propan-2-ol into a solution of **36** and C_{60} in *o*-dichlorobenzene. The X-ray struc-

[56] C.D. Gutsche, *Calixarenes*, Royal Society of Chemistry, Cambridge, UK, 1989; C.D. Gutsche, *Calixarenes Revisited*, Royal Society of Chemistry, Cambridge, UK, 1998.

[57] A. McKervey and V. Böhmer, *Chem. Br.*, 1992, 724.

[58] J. Rebek, *Chem. Soc. Rev.*, 1996, **25**, 255.

[59] K. Koh, K. Araki and S. Shinkai, *Tetrahedron Lett.*, 1994, **35**, 8255; R.H. Vreekamp, W. Verboom and D.H.N. Reinhoudt, *J. Org. Chem.*, 1996, **61**, 4287; R.K. Castellano and J. Rebek, *J. Am. Chem. Soc.*, 1998, **120**, 3657; R.K. Castellano, B.H. Kim and J. Rebek, *J. Am. Chem. Soc.*, 1997, **119**, 12671; R.K. Castellano, D.M. Rudkevich and J. Rebek, *Proc. Natl. Acad. Sci. USA*, 1997, **94**, 7132; R.K. Castellano, D.M. Rudkevich and J. Rebek, *J. Am. Chem. Soc.*, 1996, **118**, 10002.

[60] O. Mogck, V. Böhmer and W. Vogt, *Tetrahedron*, 1996, **52**, 8489.

[61] K.D. Shimizu and J. Rebek, *Proc. Natl. Acad. Sci. USA*, 1995, **92**, 12403; B.C. Harmann, K.D. Shimizu and J. Rebek, *Angew. Chem., Int. Ed. Engl.*, 1996, **35**, 1326; J. Kang and J. Rebek, *Nature*, 1997, **385**, 50.

[62] K.N. Rose, L.J. Barbour, G.W. Orr and J.L. Atwood, *Chem. Commun.*, 1998, 407.

ture of the product cage is also shown in Figure 3.7. Interestingly, C_{60} does not occupy the cavity of this product but was found in the lattice between the cage assemblies. It is probable that the cavity belonging to each cage incorporates several solvent molecules, these being necessary to maintain structural stability. The effective van der Waals volume of the cavity is *ca.* 230 Å^3; it is too large for a solitary C_{60} molecule to fill the void efficiently without the further encapsulation of a significant number of solvent molecules.

Even larger spherical structures based on calix[4]resorcinarenes have been assembled. The C-methylcalix[4]resorcinarene **37** has been shown to aggregate along with eight (adventitious) water molecules, to form a chiral spherical hexamer.[63] The latter is held together by 60 hydrogen bonds and is able to encapsulate guest molecules within its well-defined cavity (which has an internal volume of about 1375 Å^3). The X-ray structure of this impressive assembly has been determined and likened to that of a spherical virus.[64] Its overall topology conforms to that of a snub cube – one of the 13 Archimedeon solids.

37

The dimerisation of larger calix[6]arenes bearing three carboxylic acid groups in alternate positions on the upper rim has been reported. In this case the cavity is of sufficient size to encapsulate simple pyridinium salts.[37,65]

Evidence has been presented that molecular encapsulation of the type discussed in this section is to a large degree controlled by the respective volumes of the guest and the cavity in the host.[48] Thus binding of a guest molecule in the internal cavity of a receptor in solution can be anticipated when the *packing coefficient*, the ratio of the guest volume to the host volume, falls within the range 0.55 ± 0.09.

[63] L.R. MacGillivray and J.L. Atwood, *Nature*, 1997, **389**, 469.
[64] D.L.O. Caspar and A. Klug, *Cold Spring Harb. Symp. Quant. Biol.* 1962, **27**, 1.
[65] A. Arduini, L. Domiano, L. Ogliosi, A. Pochini, A. Secchi and R. Ungaro, *J. Org. Chem.*, 1997, **62**, 7866.

However, packing coefficients of up to 0.70 can be accommodated if the resulting complex is stabilised by strong intermolecular forces such as hydrogen bonds. Interestingly, the above rule also applies when more than one guest molecule is encapsulated in the receptor's cavity.

3.6   Self-replicating Systems

The ability of nucleic acids to act as templates for self-replication is a fundamental process in Nature's chemistry. The Rebek group have employed both templating and recognition effects for the production of assembled systems that promote replication of the templating molecule.[66,67] An important feature of this work is the fact that the presence of the usual weak intermolecular forces allowed the corresponding host–guest complexes to form and dissipate rapidly. The resulting dynamic behaviour provides an environment for an efficient autocatalytic replication process to occur.

A schematic representation of the proposed mechanism for replication is shown in Figure 3.8. In this, two complementary precursors (A and B) react intermolecularly (and covalently) to form the template (T). Owing to the self-complementary nature of this product and the reactants, two further molecules of A and B are able to form a ternary complex with the template. Intermolecular reaction between A and B then occurs within the complex with, finally, the weak intermolecular forces present allowing dissociation of the dimeric product. As a consequence, this leads to an increase in template concentration, with the process being autocatalytic.

The initial self-replication system investigated involved recognition between adenine and an imide of Kemp's triacid *via* hydrogen bond formation.[68–70] The

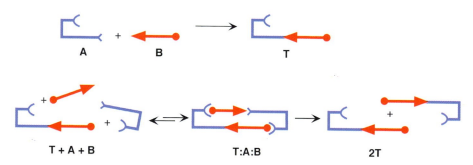

Figure 3.8 *Reaction scheme for the formation of an assembled template system for molecular self-replication*[66]

66   C. Andreu, R. Beerli, N. Branda, M. Conn, J. de Mendoza, A. Galan, I. Huc, Y. Kato, M. Tymoschenko, C. Valdéz, E. Wintner, R. Wyler and J. Rebek, *Pure Appl. Chem.*, 1993, **65**, 2313.
67   E.A. Winter, M.M. Conn and J. Rebek, *Acc. Chem. Res.*, 1994, **27**, 198.
68   T. Tjivikua, P. Ballester and J. Rebek, *J. Am. Chem. Soc.*, 1990, **112**, 1249.
69   J.S. Nowick, Q. Feng, T. Tjivikua, P. Ballester and J. Rebek, *J. Am. Chem. Soc.*, 1991, **113**, 8831.
70   M.M. Conn, E.A. Wintner and J. Rebek, *J. Am. Chem. Soc.*, 1994, **116**, 8823.

naphthoyl active ester **38** (Figure 3.9) reacts with 5′-amino-5′-deoxy-2′, 3′-isopropylideneadenosine (**39**) to form the self-complementary autocatalytic template **40**. This was proposed to be followed by the formation of the ternary complex **41** which then promotes condensation to yield a new template molecule. The autocatalytic nature of the reaction is evident from the rate acceleration caused by seeding the reaction with its product **40**.

A longer spacer group has also been inserted in the templating molecule by exchanging the 2,6-disubstituted naphthalene spacer of **38** for a 4,4′-biphenyldiyl

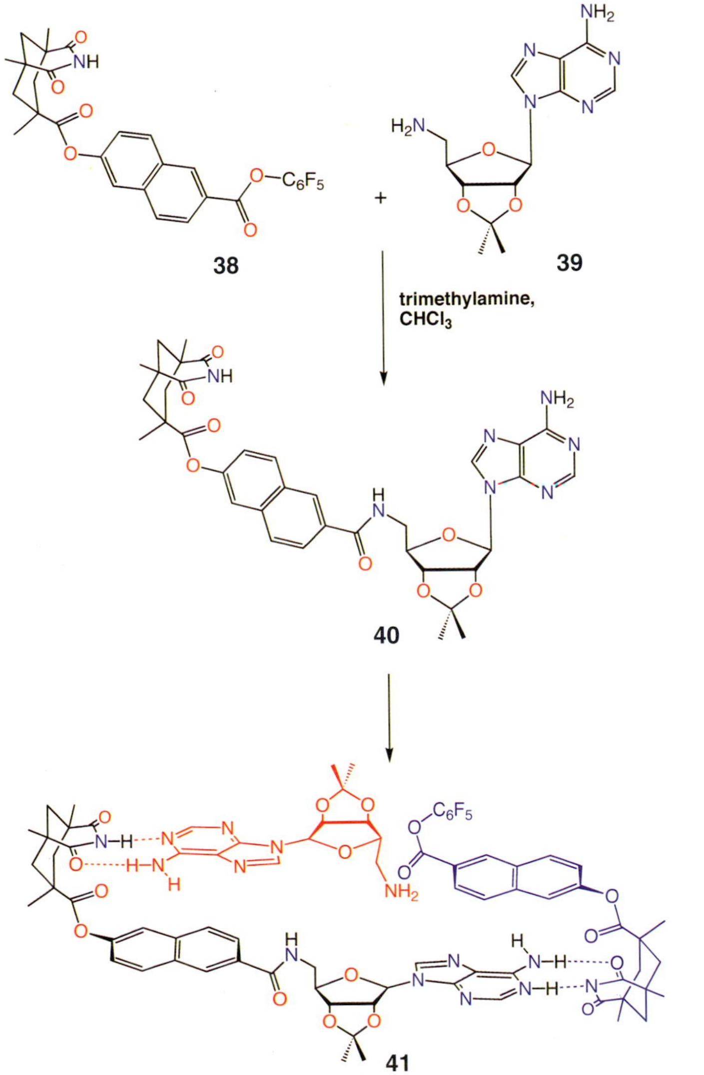

Figure 3.9 *A self-replicating system investigated by the Rebek group*[68–70]

spacer.[70,71] This resulted in a rate increase for product formation. The Rebek group has also investigated a number of other related self-replicating systems.[66,67,71]

3.7   Further Supramolecular Categories

In the following two chapters, we will explore two further major categories of discrete, non-metal-containing, supramolecular assemblies – namely, rotaxanes and the catenanes.

[71]   E.A. Wintner, M.M. Conn and J. Rebek, *J. Am. Chem. Soc.*, 1994, **116**, 8877; M.M. Conn, E.A. Wintner and J. Rebek, *Angew. Chem., Int. Ed. Engl.*, 1994, **33**, 1577; R.J. Pieters, I. Huc and J. Rebek, *Angew. Chem., Int. Ed. Engl.*, 1994, **33**, 1579.

CHAPTER 4

Rotaxanes

4.1   Introduction

Rotaxanes are molecular assemblies in which a linear component is mechanically threaded through one or more cyclic molecules – much like beads on a string – with the ends of the linear component terminated with bulky groups to prevent unthreading of the structure. The concept of a rotaxane (Latin: *rota,* wheel; *axis,* axle) is not new, with the *idea* of a rotaxane being put forward as early as 1961.[1] Following this, several early syntheses were reported in which the rotaxane was typically obtained in very low yield.[2,3]

Two extremes can be recognised in the formation of rotaxanes. First, there is 'statistical' threading which is characterised by a small K value for host–guest binding and which is usually associated with low yields of the product. Secondly, there is 'directed' or 'template' threading, characterised by larger K values, reflecting significant attraction between the components (negative ΔH values). Yields in this latter case, while often variable, tend not to be trivial and are sometimes substantial.

Our chief concern in this chapter will be with systems of this latter type. Namely, those that appear to involve a large measure of 'self-assembly' during their formation – and for which some element of the templating interactions used to direct their construction remains within the interlocked final structures. However, it is still pertinent in the present context to survey briefly a selection of both the early and more recent 'statistical' work.

In 1974, Harrison investigated the use of a statistical procedure to thread a linear methylene chain bearing large end groups, 1,10-bis(triphenylmethoxy)tridecane, through cyclic alkane rings.[4] The experiment involved a mixture of cyclic alkanes, containing all homologues from C_{14} to C_{42}, which were heated with the above linear

1   H.L. Frisch and E. Wasserman, *J. Am. Chem. Soc.,* 1961, **83**, 3789.
2   I.T. Harrison and S. Harrison, *J. Am. Chem. Soc.,* 1967, **89**, 5723; G. Schill and H. Zollenkopf, *Justus Liebigs Ann. Chem.,* 1969, **721**, 53; G. Schill and R. Henschel, *Justus Liebigs Ann. Chem.,* 1970, **731**, 113; G. Schill and H. Neubauer, *Justus Liebigs Ann. Chem.,* 1971, **750**, 76; G. Schill, C. Zürcher and W. Vetter, *Chem. Ber.,* 1973, **106**, 228.
3   G. Schill, *Catenanes, Rotaxanes and Knots,* Academic Press, New York, 1971.
4   I.T. Harrison, *J. Chem. Soc., Perkin Trans. 1,* 1974, 301.

component in the presence of naphthalene-β-sulfonic acid. The role of the acid was to promote reversible cleavage of the triaryl ether end groups to allow the 'threading' to occur before 'restoppering' the linear component. After equilibrium had been reached, the threaded products were separated by chromatography. Very small yields of threaded products were observed – ranging from 0.0013% for the C_{24}-ring to 1.6% for the C_{33}-ring, with a zero yield for the larger ring products. It appears that rings smaller than C_{22} are too small to allow threading of an alkane chain, whereas rings consisting of 34 or more methylenes permit passage of the blocking group through their respective cavities, yielding only transient formation of the corresponding rotaxane.

Interestingly, in a prior study the 29-membered ring rotaxane was shown to form selectively in low yield on heating a solution of the corresponding macrocycle directly with a C_{10}-linear component 'stoppered' with triphenylmethoxy groups.[5] Apparently this stopper is small enough to 'slip' through the macrocyclic ring at elevated temperatures. Later, Schill *et al.*[6] used a statistical procedure to obtain rotaxanes whose yields increased as a function of ring size (from 21–29 members) and chain length (from 10–38 members).

4.2 Directed Template Synthesis – Simple Host–Guest Adducts

Much of the work performed so far on larger molecular assemblies has been guided by lessons learnt from investigating smaller model species. This is well exemplified by studies undertaken by the Stoddart group.[7] In an initial investigation, the above group undertook a search for a selective synthetic receptor for the bipyridinium herbicide paraquat (**1**).[8] After a number of disappointments, they discovered that bis-*p*-phenylene-34-crown-10 (**2**) forms a 1 : 1 complex with this dicationic species. The X-ray structure of the product (Figure 4.1) revealed an interesting arrangement in which the paraquat is held centrally in the crown macrocycle such

1; paraquat   2; bis-*p*-phenylene-34-crown-10

5   I.T. Harrison, *J. Chem. Soc., Chem. Commun.*, 1972, 231.
6   G. Schill, W. Beckmann, N. Schweickert and H. Fritz, *Chem. Ber.*, 1986, **119**, 2647.
7   D.B. Amabilino and J.F. Stoddart, *New Scientist*, 1994, 25; D.B. Amabilino, J.F. Stoddart and D.J. Williams, *Chem. Mater.*, 1994, **6**, 1159; M. Belohradsky, F.M. Raymo and J.F. Stoddart, *Czech. Chem. Commun.*, 1996, **61**, 1; S.J. Stoddart, S.J. Langford and J.F. Stoddart, *Pure Appl. Chem.*, 1996, **68**, 1255; D. Philp and J.F. Stoddart, *Angew. Chem., Int. Ed. Engl.*, 1996, **35**, 1155; P.T. Glink and J.F. Stoddart, *Pure Appl. Chem.*, 1998, **70**, 419.
8   J.F. Stoddart, *Pure Appl. Chem.*, 1988, **60**, 467.

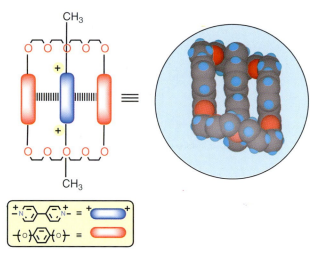

Figure 4.1 1 : 1 *Adduct formed between paraquat and bis-p-phenylene-34-crown-10*[9]

that the methyl groups of this guest protrude on either side of the ring;[9] a centrosymmetric geometry is maintained. The pair of π-electron-rich hydroquinone derivatives has the π-electron-deficient pyridinium rings of paraquat slotted between them to form a 'π-stack', with the separation between each bipyridinium ring and the hydroquinone derivative being 3.7 Å. The deep red colour of the complex, both in the solid and in solution, suggests the presence of significant charge-transfer interaction between host and guest. Other dispersive and electrostatic interactions, including weak [C–H⋯O] hydrogen bonds between the acidic hydrogen atoms α to nitrogen on the dication and ether oxygen atoms of the polyether bridge in the crown, also appear to be present. The sum of these various interactions amounts to a binding energy ($-\Delta G$) of 16.5 kJ mol^{-1} for the adduct in acetone at room temperature.

Each of the related dications **3**, **4** and **5** also forms a 1 : 1 complex with bis-*p*-phenylene-34-crown-10 (**2**).[10] X-Ray structure analyses of all three complexes (as their dihexafluorophosphate salts) were obtained (Figure 4.2). In the latter two adducts, the respective dications are again threaded through the crown in a symmetrical fashion, with the crown adopting an essentially identical conformation in both structures. Each pair of 'aromatic' OCH$_2$ groups is oriented *syn*. However, the structure obtained for the adduct incorporating **3** is different. It has the aromatic OCH$_2$ groups arranged *anti*, with a corresponding distortion of the polyether ring being present. In this case the bipyridinium dication lies almost fully within the macrocyclic ring, with two strong (almost linear) hydrogen bonds occurring to the central oxygen atom of each polyether bridge. In all three structures, π-interactions

9   P.L. Anelli, P.R. Ashton, R. Ballardini, V. Balzani, M. Delgado, M.T. Gandolfi, T.T. Goodnow, A.E. Kaifer, D. Philp, M. Pietraszkiewicz, L. Prodi, M.V. Reddington, A.M.Z. Slawin, N. Spencer, J.F. Stoddart, C. Vicent and D.J. Williams, *J. Am. Chem. Soc.*, 1992, **114**, 193.
10   P.R. Ashton, D. Philp, M.V. Reddington, A.M.Z. Slawin, N. Spencer, J.F. Stoddart and D.J. Williams, *J. Chem. Soc., Chem. Commun.*, 1991, 1680.

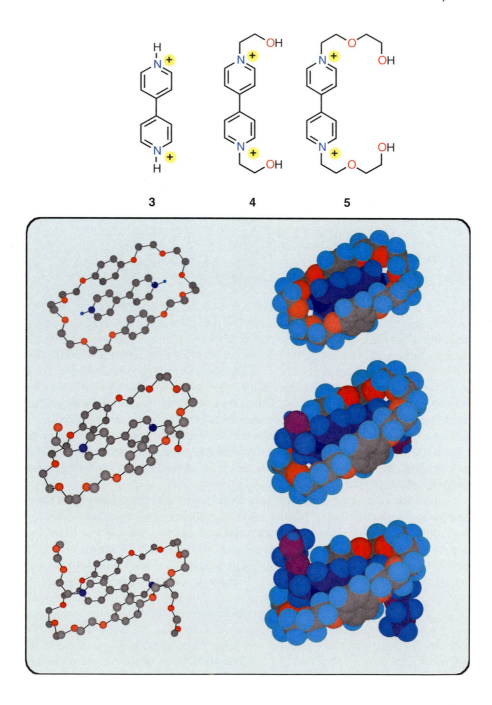

3   **4**   **5**

Figure 4.2 *X-Ray structures of the adducts formed between* **3**, **4** *and* **5** *and macrocycle* **2**[10]

between host and guest are also much in evidence. Prior NMR studies suggest that structural behaviour similar to the above is maintained in deutero-acetone. Based on these studies, Shen *et al.*[11] also synthesised the corresponding host–guest species in which the paraquat dication incorporates appended carboxylic acid or ester groups.

In a separate study, the Stoddart group set out to design a synthetic receptor for a neutral substrate such as 1,4-dimethoxybenzene (**6**) which would promote the essential threading step necessary for rotaxane formation.[12] In these investigations the host–guest roles employed in the first stage of the rotaxane synthesis as discussed above needed to be reversed – the guest is now the π-electron-rich entity. It was clear from prior molecular modelling that if two *p*-xylylene groups were used to bridge two paraquat units to pre-form the tetracationic cyclophane **7**, then this product would have its bipyridinium rings held rigidly apart at about 7 Å. It was also clear that this is a suitable dimension to accommodate a small aromatic guest (such as 1,4-dimethoxybenzene) in a π-bonded sandwich. The synthesis of the cyclic host proved to be no easy task but perseverance won in the end.

Reaction of 1,4-dimethoxybenzene with the tetracationic macrocycle led to isolation of the required 1 : 1 complex; the X-ray structure of this adduct (Figure 4.3), with the guest oriented across the macrocyclic cavity, was as predicted by

[11] Y.X. Shen, P.T. Engen, M.A.G. Berg, J.S. Merola and H.W. Gibson, *Macromolecules*, 1992, **25**, 2786.
[12] B. Odell, M.V. Reddington, A.M.Z. Slawin, N. Spencer, J.F. Stoddart and D.J. Williams, *Angew. Chem., Int. Ed. Engl.*, 1988, **27**, 1547.

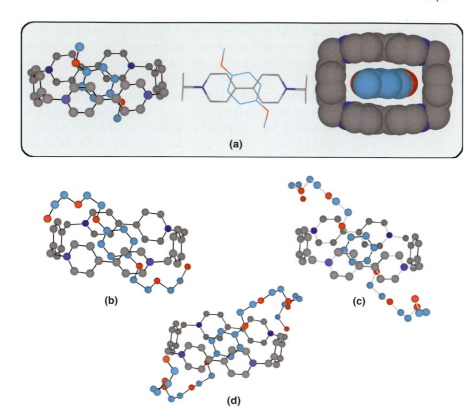

Figure 4.3 *Rotaxanes formed between macrocycle* **7** *and guests* (a) **6**, (b) **8**, (c) **9** *and*
(d) **10**[9,13]

molecular modelling.[13] Following directly from this, related complexes with **8–10** were obtained in which the hydroquinone derivatives in the molecular threads occupy the macrocyclic cavity – with the polyether tails directed over two diametrically opposite corners of the macrocyclic host.[9] The time-averaged NMR spectrum for each of the above 1 : 1 complexes indicates a structure similar to that observed in the solid state. The most significant chemical shift variations occur for the β-protons on the bipyridinium rings and for the *p*-phenylene protons. In each spectrum, the upfield shifts of the former resonance and the downfield shift of the latter are in accordance with the expected orientation of the hydroquinone ring inside the tetracationic macrocyclic cavity.

The general importance of hydrogen bonds in influencing the complex stabilities of a range of (derivatised) host–guest systems of the above type has since been documented.[14] Other studies using similar procedures resulted in the isolation of

13   P.R. Ashton, B. Odell, M.V. Reddington, A.M.Z. Slawin, J.F. Stoddart and D.J. Williams, *Angew. Chem., Int. Ed. Engl.*, 1988, **27**, 1550.
14   M. Asakawa, C.L. Brown, S. Menzer, F.M. Raymo, J.F. Stoddart and D.J. Williams, *J. Am. Chem. Soc.*, 1997, **119**, 2614.

1 : 1 and 2 : 1 complexes between the cyclobis(paraquat-*p*-phenylene) tetracation and **11** and **12**, respectively.[15] The X-ray structures of both these complexes confirm that each of the naphtho units lies within the centre of a tetracationic cyclophane. In particular, the successful preparation of the second compound conjured up the prospect of self-assembling structures with a polyrotaxane-like nature. That is, systems that would act as a prototype for the construction of a 'molecular abacus'.

OMe

OMe

11

OH

OH

12

4.3   Pseudorotaxanes

The successful syntheses of the above host–guest assemblies pointed the way to the formation of molecular threads – namely, compounds formed by threading a cyclic species, such as the cyclobis(paraquat-*p*-phenylene) tetracation, on to a separate *extended* linear component; the product is a pseudorotaxane. Pseudorotaxanes are a derivative category of the rotaxanes in which one of the assembled components is a longer (often oligomeric) linear fragment. They are thus similar to rotaxanes, with the exception that the linear component is not terminated by bulky 'stoppers'.

Based on the experiments discussed earlier, the 'directed' self-assembly of the [2]-pseudorotaxane (Figure 4.4) from the linear polyether **13** (incorporating two tetraethyleneglycol units alternating between three hydroquinone derivatives) and the corresponding tetracationic cyclophane **7** was carried out and the product was isolated.[16] As might be expected, the structure of this 1 : 1 product displays a large degree of self-ordering in the solid state, largely centred around π–π stacking.

Figure 4.4 *X-Ray structure of the [2]-pseudorotaxane formed between* **7** *and* **13**[16]

15   M.V. Reddington, A.M.Z. Slawin, N. Spencer, J.F. Stoddart, C. Vicent and D.J. Williams, *J. Chem. Soc., Chem. Commun.*, 1991, 630.
16   P.L. Anelli, P.R. Ashton, N. Spencer, A.M.Z. Slawin, J.F. Stoddart and D.J. Williams, *Angew. Chem., Int. Ed. Engl.*, 1991, **30**, 1036.

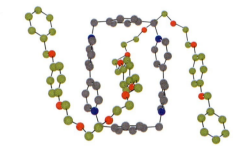

13

The generality of this ordering process was established by the further syntheses of the [2]-pseudorotaxane shown in Figure 4.5[17] and the (then) novel [3]-pseudorotaxane shown in Figure 4.6.[17] The first of these species involves the 1 : 1 self-assembly of **14** with one molecule of the bis(paraquat-*p*-phenylene) tetracationic macrocycle, while in the second species the polyether component is **15** and the corresponding ratio is 1 : 2. The successful construction of the latter species represented somewhat of a landmark in the use of self-assembly processes of the present type since it demonstrated the feasibility of extending the work to large synthetic

Figure 4.5 *X-Ray structure of the [2]-pseudorotaxane formed between* **7** *and* **14**[16]

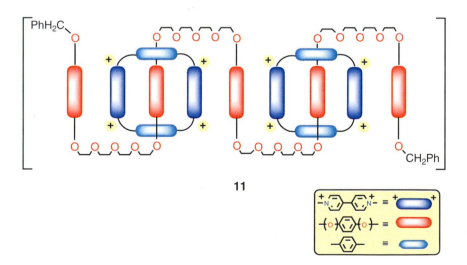

11

Figure 4.6 *Schematic of the [3]-pseudorotaxane* **11** *formed between* **7** *and* **15**[17]

[17]   P.R. Ashton, D. Philp, N. Spencer and J.F. Stoddart, *J. Chem. Soc., Chem. Commun.*, 1991, 1677.

14

15

assemblies involving more than two components. NMR evidence again supported the formation of a tightly-packed π–π stacked structure, which, in this case, bears an eight positive charge. Such a structure is of nanometre dimensions: approximately 30 Å long and 15 Å across. In broad terms, the study was viewed as a model for the construction of large polymolecular arrays of the type that one day may act as information storing and processing devices. In view of this, attention was given by the Stoddart group to the nature of both constitutional and translational isomerism in [n]-pseudorotaxanes of this general type.[18]

While the above studies are representative of the area, it should be noted that a considerable number of other [2]-pseudorotaxanes have been investigated. These include a system exhibiting enantioselective self-assembly during its formation[19] as well as systems incorporating a new structural motif involving a 1,2-bis(pyridinium)ethane 'axle' and a 24-crown-8 ether 'wheel'.[20] A number of poly-pseudorotaxanes (and related poly-rotaxanes)[21,22] as well as constitutionally asymmetric and chiral [2]-pseudorotaxanes[23] have been reported. New, multiply-stranded and multiply-encircled systems have also been prepared.[24] The solid state structure of

[18] D.B. Amabilino, P.-L. Anelli, P.R. Ashton, G.R. Brown, E. Córdova, L.A. Godinez, W. Hayes, A.E. Kaifer, D. Philp A.M.Z. Slawin, N. Spencer, J.F. Stoddart, M.S. Tolley and D.J. Williams, *J. Am. Chem. Soc.*, 1995, **117**, 11142.

[19] M. Asakawa, H.M. Janssen, E.W. Meijer, D. Pasini and J.F. Stoddart, *Eur. J. Org. Chem.*, 1998, 983.

[20] S.J. Loeb and J.A. Wisner, *Angew. Chem., Int. Ed.,* 1998, **37**, 2838.

[21] D.B. Amabilino, I.W. Parsons and J.F. Stoddart, *Trends Polym. Sci.*, 1994, **2**, 146; Y.X. Shen, D. Xie and H.W. Gibson, *J. Am. Chem. Soc.*, 1994, **116**, 537 and references therein.

[22] P.E. Mason, I.W. Parsons and M.S. Tolley, *Angew. Chem., Int. Ed. Engl.*, 1996, **35**, 2238; H.W. Gibson, D.S. Nagvekar, J. Powell, C.G. Gong and W.S. Bryant, *Tetrahedron*, 1997, **53**, 15197.

[23] M. Asakawa, P.R. Ashton, W. Hayes, H.M. Janssen, E.W. Meijer, S. Menzer, D. Pasini, J.F. Stoddart, A.J.P. White and D.J. Williams, *J. Am. Chem. Soc.*, 1998, **120**, 920.

[24] P.R. Ashton, M.C.T. Fyfe, P.T. Glink, S. Menzer, J.F. Stoddart, A.J.P. White and D.J. Williams, *J. Am. Chem. Soc.*, 1997, **119**, 12514.

the [2]-pseudorotaxane analogue of the structure shown in Figure 4.5, in which all three of the hydroquinone derivatives are replaced by 1,5-dioxynaphthalene groups, has been determined.[25] Extensive self-organisation occurs on crystallisation in this case such that a highly-ordered, layered superstructure of [2]-pseudorotaxane molecules is formed. The two-dimensional framework of π-acceptor and π-donor units that assembles is largely held together by arrays of both face-to-face and face-to-edge π-interactions.

Consideration has been given to the potential use of pseudorotaxanes in molecular devices. Ultimately, a molecular device such as a switch, needs to be addressable by means of an external stimulus such as light (which also provides the energy necessary for its operation).[26] 'Switching' experiments have been conducted using the 1 : 1 pseudorotaxane formed between the linear thread **16**, incorporating a 1,5-dioxynaphthalene group, and the usual cyclophane tetracation employed by the Stoddart group in prior studies.[27] For this system, the threading (and unthreading – see below) of **16** can be followed by absorption and emission spectra, based

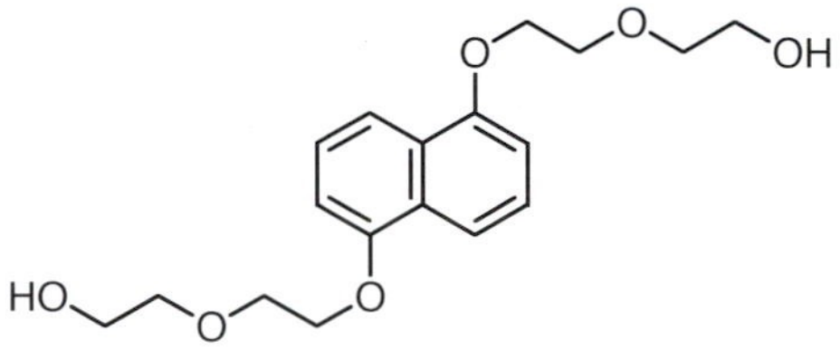

16

on the formation of a charge-transfer band in the visible region on complexation and the presence of strong luminescence arising from the 1,5-dioxynaphthalene linear fragment when uncomplexed. The first excited state of 9-anthracenecarboxylic acid is a well-known strong reductant with a long half-life. When a solution of the above pseudorotaxane components in water (at 6.0×10^{-5} mol dm^{-3}) was treated with this acid and the solution irradiated, reduction of the π-electron-deficient paraquat units in tetracationic cyclophane occurred. This, in turn, favours the unthreading of the host–guest system – but unfortunately the process is slow and it does not compete with the faster back-electron-transfer from the reduced [2]-pseudocatenane to the oxidised form of the 9-anthrocenecarboxylic acid. However, the addition of a sacrificial quantity of triethanolamine as a reductant results in the oxidised form being rapidly scavenged, preventing the above 'back'-electron-transfer reaction. Hence the pseudorotaxane remains reduced, removing the charge-transfer attraction between host and guest, and unthreading occurs. After 25 minutes of irradiation, 35% of the original pseudorotaxane was found to be unthreaded.

[25] P.R. Ashton, D. Philp, N. Spencer, J.F. Stoddart and D.J. Williams, *J. Chem. Soc., Chem. Commun.*, 1994, 181.

[26] P.R. Ashton, R. Ballardini, V. Balzani, S.E. Boyd, A. Credi, M.T. Gandolfi, M. Gomez-Lopez, S. Iqbal, D. Philp, J.A. Preece, L. Prodi, H.G. Ricketts, J.F. Stoddart, M.S. Tolley, M. Venturi, A.J. White and J.P. Williams, *Chem. Eur. J.*, 1997, **3**, 152.

[27] R. Ballardini, V. Balzani, M.T. Gandolfi, L. Prodi, M. Venturi, D. Philp, H.G. Ricketts and J.F. Stoddart, *Angew. Chem., Int. Ed. Engl.*, 1993, **32**, 1301.

Tetrathiafulvalene-containing systems have also been reported.[28,29] One such pseudorotaxane represents a redox-active system in which a similar molecular motion can be controlled by two different inputs.[29] Reversible dethreading–rethreading cycles involving this product can be performed by *either* oxidation and consecutive reduction of the linear component *or* reduction and consecutive oxidation of the electron-accepting cyclophane tetracation. As such, the system corresponds to a further level of sophistication in systems of this type. Formally, the input (electrochemical)/output (absorption spectrum) characteristics of this assembly correspond to those of an XNOR logic gate.

Mention also needs to be made of a number of poly-pseudorotaxane systems whose syntheses were claimed largely to involve the 'statistical' threading technique. In one such study, a mixture of dibenzo-crown ethers (that ranged upwards in size from dibenzo-30-crown-10) was treated with polyethylene glycol chains of different length at 120 °C.[30] The products from these experiments were then 'frozen' by reaction of the free alcohol groups with naphthalene-1,5-diisocyanate (to yield high molecular weight polyurethanes). Although it was realised that the naphthalene group is not bulky enough to prevent unthreading of the cyclic crown ethers, such a process was only expected to occur for the relatively small number of rings that were positioned near the end of the polymer chain. In any case, cooling and storing the product in the solid state was found to minimise such loss. Analysis of the reaction products enabled the degree of 'threading' to be established; it also allowed a detailed analysis of the factors influencing the efficiency of the threading process to be assessed. By variation of the conditions, moderate yields (up to 15%) were achieved for individual experiments. Such yields, together with the relative stabilities of the intertwined products, were in part accounted for by the presence of dipole–dipole interactions between the bis-methylenedioxy units present in both the cyclic and threaded components.

The statistical threading concept has also been extended to the synthesis of polyurethane pseudorotaxanes based on 60-crown-20 and 36-crown-12. In these cases the polymerisation of tetraethylene glycol and methylene di-*p*-phenyl diisocyanate was performed using the respective melted crowns as the solvent.[31] Threading efficiencies of 57 and 16%, respectively, were obtained. These values were postulated to reflect, in part, the general compatibility between the crown ethers and the glycol moieties. The use of rather large crowns probably also aids the threading process in these systems.

The chemically controlled unthreading of particular [2]-pseudorotaxanes has been investigated.[32,33] One system employed for a study of this type was composed of a

28 Z.-T. Li, P.C. Stein, J. Becher, D. Jensen, P. Mork and N. Svenstrup, *Chem. Eur. J.*, 1996, **2**, 624.
29 M. Asakawa, P.R. Ashton, V. Balzani, A. Credi, G. Mattersteig, O.A. Matthews, M. Montalti, N. Spencer, J.F. Stoddart and M. Venturi, *Chem. Eur. J.*, 1997, **3**, 1992.
30 G. Agam, D. Graiver and A. Zilkha, *J. Am. Chem. Soc.*, 1976, **98**, 5206.
31 Y.X. Shen and H.W. Gibson, *Macromolecules*, 1992, **25**, 2058.
32 M. Asakawa, S. Iqbal, J.F. Stoddart and N.D. Tinker, *Angew. Chem., Int. Ed. Engl.*, 1996, **35**, 976; R. Ballardini, V. Balzani, A. Credi, M.T. Gandolfi, S.J. Langford, S. Menzer, L. Prodi, J.F. Stoddart, M. Venturi and D.J. Williams, *Angew. Chem., Int. Ed. Engl.*, 1996, **35**, 978; A. Credi, V. Balzani, S.J. Langford and J.F. Stoddart, *J. Am. Chem. Soc.*, 1997, **119**, 2679.
33 P.R. Ashton, S. Iqbal, J.F. Stoddart and N.D. Tinker, *Chem. Commun.*, 1996, 479.

cyclobis(paraquat-*p*-phenylene) tetracation surrounding a polyether chain inter-
cepted in its middle by a hydroquinone ring and terminated at each end by a
12-crown-4 macrocycle.[33] Dethreading of this species has been demonstrated to
occur in acetonitrile on addition of alkali metal salts. The latter bind at the crown
polyether sites, apparently causing the disassembly of the structure through
unfavourable electrostatic interaction of the metal centres with the tetracationic
paraquat-ring derivative.

4.4   Simple Charged Rotaxanes

Of course, it was clear from the beginning that a [2]-rotaxane could be constructed either by threading the macrocycle onto the linear component then stoppering
it or, alternatively, by starting with the stoppered linear fragment then clipping
together the macrocyclic component around it (Figure 4.7).[34] Both approaches have
been employed successfully, for example, to obtain the single hydroquinone-containing [2]-rotaxane whose solid state structure is illustrated in Figure 4.8.[9]

   The realisation that the polyether chains attached to the hydroquinone fragments
in systems such as these must be at least diethyleneglycol (or longer) units also
served to confirm the finely balanced nature of the assembly process.

   As in the above system, the triisopropylsilyl group has proved to be a convenient stopper for the ends of polyether chains incorporating hydroquinone derivatives even though other bulky groups, such as adamantoyl,[35] have also been used.
Indeed, by fine-tuning the size of the stoppers, in specific instances the cyclic com-

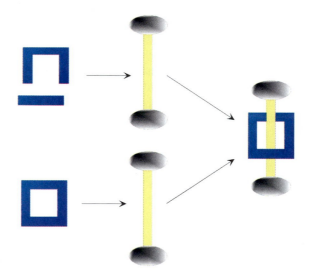

Figure 4.7 *Alternative strategies for the assembly of rotaxane*[34]

[34]   J.F. Stoddart, *Chem. Br.*, 1991, 714.
[35]   P.R. Ashton, M. Grognuz, A.M.Z. Slawin, J.F. Stoddart and D.J. Williams, *Tetrahedron Lett.*, 1991,
        32, 6235.

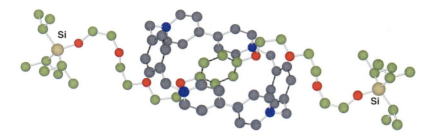

Figure 4.8 *X-Ray structure of a single hydroquinone-containing [2]-rotaxane synthesised by both strategies outlined in Figure 4.7[9]*

ponent can be 'slipped' over them at elevated temperatures but still be effectively held in place when the solution temperature is lowered. 'Slippage' (see Figure 4.9) thus corresponds to a third strategy for synthesising rotaxanes.[36,37] Further examples illustrating this process are presented in Section 4.5 of this chapter.

Success in preparing [2]-rotaxanes by both 'threading' and 'clipping' gave rise to the prospect of using a strongly bound uncharged guest, such as **17**, as a template to produce the tetracationic cyclophane **7** by the clipping mechanism then isolating it free of the linear component by use of a suitable unthreading procedure.[9] For example, using **17**, cyclisation proceeds smoothly at room temperature to yield the highly coloured 1 : 1 complex (Figure 4.10). Continuous extraction of an aqueous solution of this product with dichloromethane readily separates the two components. The template moiety **17** may be recovered from the dichloromethane phase while the charged cyclophane **18** remains in the water phase. By adjusting the conditions for the cyclisation reaction (Figure 4.10), the cyclophane tetracation was obtained in a final yield of 62%.

The first [2]-rotaxane prepared by the Stoddart group to act as a molecular shuttle,[38] **19**, is shown in Figure 4.11. Formation of this system involved the clipping

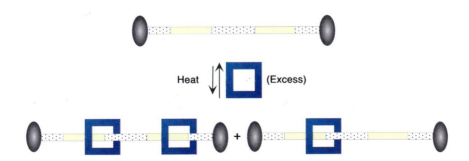

Figure 4.9 *Self-assembly of [2]- and [3]-rotaxanes by slippage[36]*

[36] D.B. Amabilino and J.F. Stoddart, *Pure Appl. Chem.*, 1993, **65**, 2351.
[37] F.M. Raymo and J.F. Stoddart, *Pure Appl. Chem.*, 1997, **69**, 1987.
[38] P.L. Anelli, N. Spencer and J.F. Stoddart, *J. Am. Chem. Soc.*, 1991, **113**, 5131.

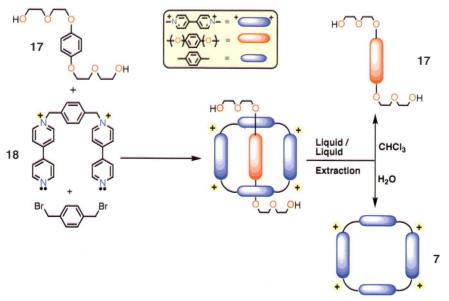

Figure 4.10 *The template synthesis of* **7** *using guest* **17**[9]

together of the cyclobis(paraquat-*p*-phenylene) tetracation (once again, from its tricationic precursor and *p*-dibromoxylene) around the capped polyether, bis(silyloxyethoxyethoxyethoxyphenoxyethoxyethoxy)ether. The dynamic ¹H NMR behaviour of this deep-orange product in acetone-d₆ indicated that the cyclophane 'bead' shuttles to and fro between the two hydroquinone derivative sites at a rate of approximately 500 times a second at room temperature (see Figure 4.11).[34] As expected,

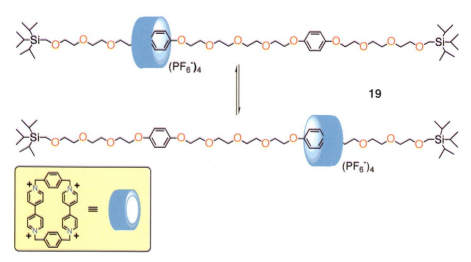

Figure 4.11 *A 'molecular shuttle' – the cyclophane 'bead' exchanges between the hydroquinone sites at approximately 500 Hz at room temperature*[38]

under conditions of slow exchange (–50 °C), separate signals are observed for the two hydroquinone rings (occupied and unoccupied) as well as for the triisopropyl groups of the silyl 'stoppers' which are now no longer equivalent.

In order to assess the possibility of including some addressable functionality into [2]-rotaxanes of this type, the analogue of **19**, in which the triisopropylsilyl stoppers were replaced with zinc tetraarylporphyrin groups, was prepared.[39] The similar system incorporating non-metallated tetraarylporphyrin groups and one hydroquinone derivative ring was also synthesised. The concept was to use end groups, which could serve two roles – they would be large enough to act as 'stoppers', and might also act as photoactive sites. As mentioned already, it was considered that incorporation of a photoactive site in a rotaxane might ultimately provide the means of inducing light-controlled molecular switching; however, the arrangement in the present systems is, of course, not suitable for this purpose.

An extension of the above studies has resulted in the synthesis of the homologous series of [2]-rotaxanes with $n = 2$–5 shown in Figure 4.12.[40] These were prepared using the 'clipping' procedure starting from the corresponding series of linear polyethers incorporating between 2 and 5 hydroquinone derivative residues (which were 'pre-stoppered' by bulky adamantoylcarbonyl terminal groups). It is of interest that the yields of the respective [2]-catenanes increase as the number of hydroquinone derivative groups in the linear component increases from 2 to 5 (yields: 1, 25, 29 and 40%, respectively).

Preliminary investigations suggested that it would be possible to form a related poly-pseudorotaxane built up from a linear component incorporating 'linked' linear polyether strings.[43] Thus, reaction of the linear component incorporating three hydroquinone residues with methylenebis(4-phenylisocyanate) yielded the

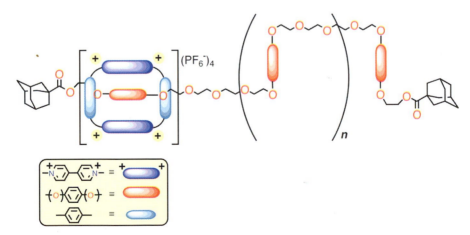

Figure 4.12 *A homologous series of rotaxanes* (n = 2–5), *prepared by 'clipping'*[40]

39 P.R. Ashton, M.R. Johnston, J.F. Stoddart, M.S. Tolley and J.W. Wheeler, *J. Chem. Soc., Chem. Commun.*, 1992, 1128.
40 X. Sun, D.B. Amabilino, I.W. Parsons and J.F. Stoddart, *Polymer Preprints*, 1993, **34**, 104.

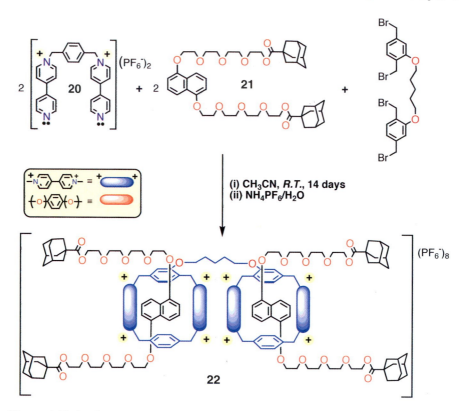

Figure 4.13 *Synthesis of the rotaxane* **22** *via a double clipping mechanism*[41]

corresponding polyurethane containing about nine repeating units. Reaction of this linear component with excess of the cyclobis(paraquat-*p*-phenylene)cyclophane tetracation results in the generation of a deep-orange product, strongly suggesting that the self-assembly process had taken place.

Rotaxane entities bearing other molecular architectures have also been synthesised.[41,42] For example, the clipping mechanism has been employed to clip the 'linked' reagent **20** around two dumb-bell-shaped components of type **21**, giving rise to the bis-[2]-rotaxane **22** shown in Figure 4.13.

A template-directed synthesis of [2]-rotaxanes, with a yield in one case of 72%, has been reported. The experiments employed dumb-bell-shaped components incorporating terminal triisopropylsilyl stoppers connected to a central 1,5-dioxynaphthalene recognition site by [–CH₂CH₂O–]ₙ spacers (*n* = 1–3).[43] These components were used as templates for the synthesis ('clipping reaction') of the corresponding rotaxanes incorporating cyclobis(paraquat-*p*-phenylene) as the ring component. The

41 P.R. Ashton, J.A. Preece, J.F. Stoddart, M.S. Tolley, *Synlett.*, 1994, 789.
42 P.R. Ashton, J. Huff, S. Menzer, I.W. Parsons, J.A. Preece, J.F. Stoddart, M.S. Tolley, A.J.P. White and D.J. Williams, *Chem. Eur. J.*, 1996, **2**, 31.
43 J.A. Bravo, F.M. Raymo, J.F. Stoddart, A.J.P. White and D.J. Williams, *Eur. J. Org. Chem.*, 1998, 2565.

length of the polyether chain influences the efficiency of the template procedure. Rotaxane formation occurs only if $n > 1$; when $n = 3$, the high yield (72%) mentioned above was achieved.

A 'reverse' strategy for construction of a [2]-rotaxane has also proved successful.[44] For this, the cyclic component was the crown compound, bis-*p*-phenylene-34-crown-10, while the 'linear' component contained two π-electron-deficient 4,4'-bipyridinium groups. Thus, this approach differs from that described for rotaxane formation so far in this chapter [in which the linear component incorporated π-electron-rich aromatic rings, such as hydroquinone-derived rings, while the cyclic component was a π-electron-deficient, tetracationic cyclobis(paraquat-*p*-phenylene) ring]. The previously documented strong and selective binding of the above crown macrocycle for 4,4'-bipyridinium dications provided the starting point for the design of the new synthesis; interestingly, the corresponding monocation is not bound.

The new synthesis is summarised in Figure 4.14. Initially, the bis-monoquaternary salt **23** – which has no affinity for the crown macrocycle – was reacted with one molar equivalent of the bulky blocking group **24** in the presence of the crown macrocycle. This generates a tricationic linear species **25**, incorporating both a pyridylpyridinium monocation and a 4,4'-bipyridinium dication. The crown macrocycle then binds strongly with the latter entity to yield the intermediate

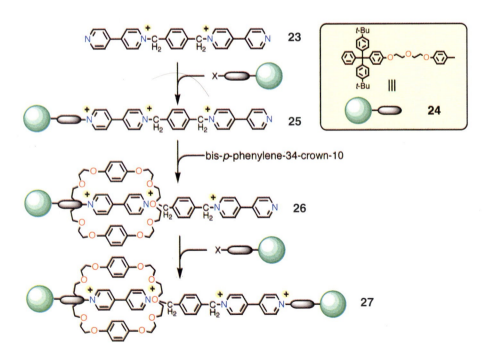

Figure 4.14 *Synthesis of the 'two site' [2]-rotaxane* **27** *by a stepwise 'stoppering' procedure*[44]

44   P.R. Ashton, D. Philp, N. Spencer and J.F. Stoddart, *J. Chem. Soc., Chem. Commun.*, 1992, 1124.

(1 : 1) complex **26**. Reaction with a further molar equivalent of the blocking group produces the required [2]-rotaxane **27**.

In the above procedure, the role of the blocking group is clearly of crucial importance. It must block one end of the linear component while simultaneously generating an extra charge on this species. The latter provides the driving force for the selective threading of one, and only one, crown macrocycle. Final reaction with a second blocking group serves to lock this ring on the 'axle'. The system is thus self-contained in that no external reagent or template is necessary to aid the self-assembly. The crown macrocycle was found to shuttle between the two charged domains with a site-exchange rate of 3×10^5 s^{-1} at 25 °C (in deutero-acetone).

The self-assembly of [2]-, [3]- and [4]-rotaxanes, with bis-*p*-phenylene-34-crown-10 as the ring component(s) encircling linear bypyridinium-based frameworks terminated with dendritic stoppers, has been reported.[45] As a consequence of the presence of the (hydrophobic) dendritic fragments, these rotaxanes are soluble in a wide range of organic solvents (in spite of the polycationic natures of the bipyridinium-containing backbones). For one [2]-rotaxane incorporating two bipyridinium sites on the linear component, the shuttling of a single macrocyclic ring between these sites has been studied in a range of solvents using ^{1}H NMR. The polarity of the medium was demonstrated to have a marked influence on the rate of shuttling; for example, on passing from deutero-chloroform to deutero-acetone, the shuttling rate was observed to increase from about 200 to about 33 000 times per second in this case.

It is noted that further reports have appeared describing other molecular shuttles exhibiting a variety of molecular architectures.[46]

Finally, other charged [2]- and [3]-rotaxanes have also been reported.[47] These incorporate a linear conjugated component formed by Glaser coupling of a (water soluble) alkyne – with the coupled product threaded through new charged cyclophanes. For these systems, the assembly process is driven by hydrophobic binding between the respective components.

4.5   Synthesis by Slippage

In other studies, matched complementarity between the cavity size of the macrocyclic component and the size of the bulky stopper groups has been exploited to allow preparation of rotaxanes by means of the 'slippage' mechanism mentioned previously.

Clearly, the tris(4-*t*-butylphenyl)methyl stopper discussed in the previous section

[45]   D.B. Amabilino, P.R. Ashton, V. Balzani, C.L. Brown, A. Credi, J.M.J. Frechet, J.W. Leon, F.M. Raymo, N. Spencer, J.F. Stoddart and M. Venturi, *J. Am. Chem. Soc.*, 1996, **118**, 12012.
[46]   See, for example: (a) R.A. Bissell, E. Córdova, A.E. Kaifer and J.F. Stoddart, *Nature*, 1994, **369**, 133; (b) H. Murakami, A. Kawabuchi, K. Kotoo, M. Kunitake and N. Nakashima, *J. Am. Chem. Soc.*, 1997, **119**, 7605 and references therein; (c) P.L. Anelli, M. Asakawa, P.R. Ashton, R.A. Bissell, G. Clavier, R. Gorski, A.E. Kaifer, S.J. Langford, G. Mattersteig, S. Menzer, D. Philp, A.M.Z. Slawin, N. Spencer, J.F. Stoddart, M.S. Tolley and D.J. Williams, *Chem. Eur. J.*, 1997, **3**, 1113 and references therein.
[47]   S. Anderson, R.T. Aplin, T.D.W. Claridge, T. Goodson, A.C. Maciel, G. Rumbles, J.F. Ryan and H. Anderson, *J. Chem. Soc., Perkin Trans. 1*, 1998, 2383.

is too large to allow passage of the usual 34-membered crown macrocycle over it. However, it was reasoned that careful (downward) size adjustment of the stoppers might enable synthesis of the corresponding rotaxane directly at elevated temperatures.[48] Such a process would be driven by thermodynamics but, once the rotaxane is formed, both a thermodynamic and kinetic barrier may operate to inhibit dissociation at ordinary temperatures. The possibility that more than one crown macrocycle might be slipped on to an appropriately designed linear component seemed also an interesting prospect.

R = H, CH₃, C₂H₅

28

In accordance with the above, three different 4,4′-bipyridinium dications of type **28** with R = H, CH₃ or C₂H₅ were prepared as their (di-)hexafluorophosphate salts. Heating of an acetonitrile solution of each of these with four mole equivalents of bis-*p*-phenylene-34-crown-10 for 10 days at 60 °C led to isolation of the corresponding rotaxanes in 52, 45 and 47% yields, respectively. When the concentration of the crown was doubled, the yield of the first product increased to 87%. Significantly, when the size of the stopper was increased such that R corresponded to *iso*-propyl, it was no longer possible to synthesise the corresponding product by this procedure at a preparatively useful rate; namely, the rate of formation was inhibited by around ten-fold.

The above results indicate that there is little difference between the rates of formation of the three rotaxanes (with R equal to H, CH₃ or C₂H₅), suggesting that the steric hindrance to slippage is similar in all three cases. When the rotaxane with R = H was heated in deutero-dimethyl sulfoxide, there was loss of the crown component from the assembly – at 60 °C, the loss was less than 25% after six hours.

Other rotaxanes have been prepared using the 'slippage' strategy.[49,50] One such example consists of a linear component (terminated by carbohydrate stoppers) that contained a hydroquinone recognition site. This was demonstrated to interact with cyclobis(paraquat-*p*-phenylene) tetrachloride in D₂O over one week at room temperature to yield the corresponding water-soluble catenane.[50]

[48] P.R. Ashton, M. Belohradsky, D. Philp and J.F. Stoddart, *J. Chem. Soc., Chem. Commun.*, 1993, 1269.

[49] P.R. Ashton, R. Ballardini, V. Balzani, M. Belohradsky, M.T. Gandolfi, D. Philp, L. Prodi, F.M. Raymo, M.V. Reddington, N. Spencer, J.F. Stoddart, M. Ventari and D.J. Williams, *J. Am. Chem. Soc.*, 1996, **118**, 4931; M. Asakawa, P.R. Ashton, S. Iqbal, A. Quick, J.F. Stoddart, N.D. Tinker, A.J.P. White and D.J. Wiliams, *Israel J. Chem.*, 1996, **36**, 329; F.M. Raymo and J.F. Stoddart, *Pure Appl. Chem.*, 1997, **69**, 1987; M. Asakawa, P.R. Ashton, R. Ballardini, V. Balzani, M. Belohradsky, M.T. Gandolfi, O. Kocian, L. Prodi, F.M. Raymo, J.F. Stoddart and M. Venturi, *J. Am. Chem. Soc.*, 1997, **119**, 302.

[50] P.R. Ashton, S.R.L. Everitt, M. Gomez-Lopez, N. Jayaraman and J.F. Stoddart, *Tetrahedron Lett.*, 1997, **38**, 5691.

The slippage procedure lends itself to the preparation of [*n*]-rotaxanes that are based on longer linear fragments incorporating more than one dicationic binding site and fitted with stoppers of appropriate size. The successful synthesis of the [3]-rotaxane **29** has been achieved by this means. This product, which is formed in good yield when the crown macrocycle is present in substantial excess, was

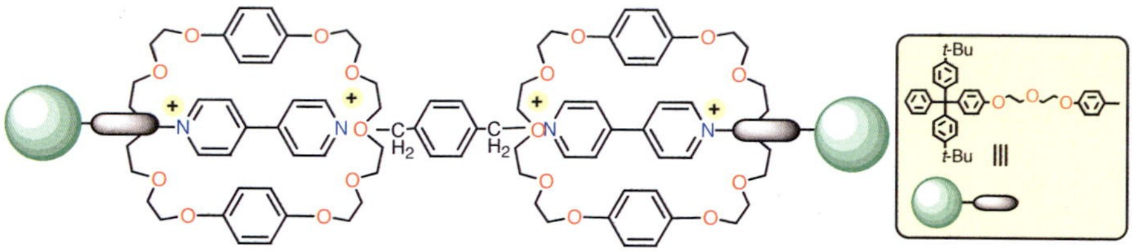

29

initially obtained from acetonitrile solution at 55 °C as a mixture with its corresponding [2]-rotaxane.[51] Both products are stable at room temperature and were separated by chromatography on silica. Clearly, slippage provides a useful technique for the preparation of oligo- and poly-rotaxanes – including ones in which many recognition sites along the linear component are encircled by *different* macrocyclic components. In this regard, self-assembly using the slippage procedure has resulted in the formation of the [3]-rotaxane **30**.[52] In this system, the stoppered

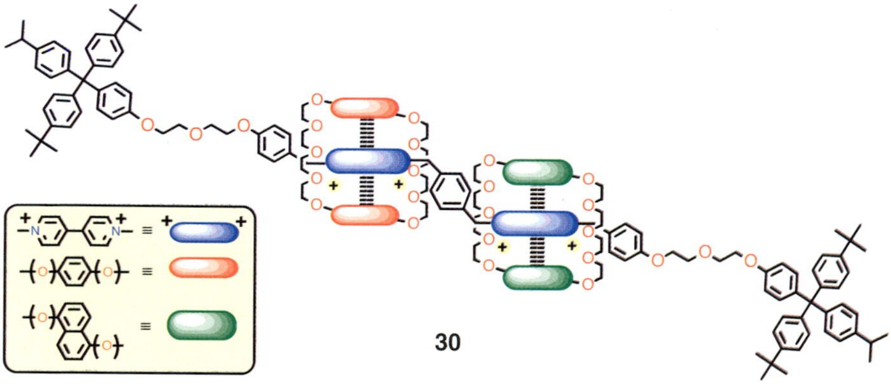

30

linear component incorporates two bipyridinium units, which are encircled respectively by a bis-*p*-phenylene-34-crown-10 macrocycle and a 1,5-dinaphtho-38-crown-10 macrocycle.

The construction of even more elaborate structures has proved possible using the slippage procedure. These include three novel rotaxanes incorporating one, two and three bis-*p*-phenylene-34-crown-10 macrocycles, respectively.[53] Each of these

51   P.R. Ashton, M. Belohradsky, D. Philp, N. Spencer and J.F. Stoddart, *J. Chem. Soc., Chem. Commun.*, 1993, 1274.

52   D.B. Amabilino, P.R. Ashton, M. Belohradsky, F.M. Raymo and J.F. Stoddart, *J. Chem. Soc., Chem. Commun.*, 1995, 747.

53   D.B. Amabilino, P.R. Ashton, M. Belohradsky, F.M. Raymo and J.F. Stoddart, *J. Chem. Soc., Chem. Commun.*, 1995, 751.

systems is based on a single branched component consisting of three bipyridinium units covalently attached to a central 1,3,5-trisubstituted benzene core; there is a substituted tetraarylmethane blocking group attached to the 'outer' edge of each of the 'anchored' bipyridinium groups. The slippage reaction was performed in acetonitrile and yielded a mixture of all three possible products, which could be readily separated using chromatography.

4.6   Towards 'Switchable' Systems

Following the successful synthesis of the range of simple 'symmetrical' rotaxanes discussed earlier, the way lay clear for the preparation of related systems in which non-identical 'stations' were incorporated in the thread. Overlaying this plan was the idea that non-identical sites could be addressed selectively by chemical, electrochemical or photochemical means such that a mechanism would be provided for shuttling the 'bead' backwards and forwards between them. Once again, such a system might be considered to be a prototype for a molecular-scale machine that is capable of receiving and storing information in a controllable manner.

In an initial study of this type,[54] it was reasoned that replacement of one of the two degenerate hydroquinone derivatives in the linear component of a [2]-rotaxane such as **19** by a group with lower π-donating ability (that is, one with a concomitant higher oxidation potential) might allow the use of electrochemistry to control the positioning of a cyclobis(paraquat-*p*-phenylene) bead. Accordingly, a *p*-xylyl residue was substituted for one of the hydroquinone derivatives. The preferential oxidation of the π-electron-rich hydroquinone-derived site in such a system should result in this site having lower affinity for the tetracationic bead – perhaps causing it to switch occupancy to the *p*-xylyl 'station', in spite of the latter's lower π-electron richness. In this manner, redox control of the shuttle seemed possible. Using the usual self-assembly synthetic procedures,[55] it proved possible to assemble the non-symmetrical [2]-rotaxane **31** (Figure 4.15). However, for the redox switching experiment to be decisive there needs to be near 100% occupancy of the hydroquinone site by the tetracationic cyclophane before the redox step. Unfortunately, this was not found to be the case – the ^{1}H NMR spectrum (slow-exchange) at 60 °C indicated 70% occupancy of the hydroquinone site and 30% of the *p*-xylyl site.

In view of this difficulty, a 'reverse' strategy was attempted. In this, instead of replacing one hydroquinone ring by a *less* π-electron-rich unit, it was replaced by a *more* π-electron-rich entity. It was reasoned that the π-donor moiety chosen as the replacement for the hydroquinone should have an oxidation potential of less than +1.0V (to form a positively charged radical). It also needed to be not too large and to be essentially non-basic. A 2,3,5-trisubstituted indole derivative was selected. Thus, it was anticipated that the highly π-electron-rich nature of the indole moiety might promote sole occupation of its site by the usual tetracationic cyclophane. It was also predicted that any change of occupancy after the oxidation

54   P.R. Ashton, R.A. Bissell, N. Spencer, J.F. Stoddart and M.S. Tolley, *Synlett*, 1992, 914.
55   P.R. Ashton, R.A. Bissell, R. Górski, D. Philp, N. Spencer, J.F. Stoddart and M.S. Tolley, *Synlett*, 1992, 919.

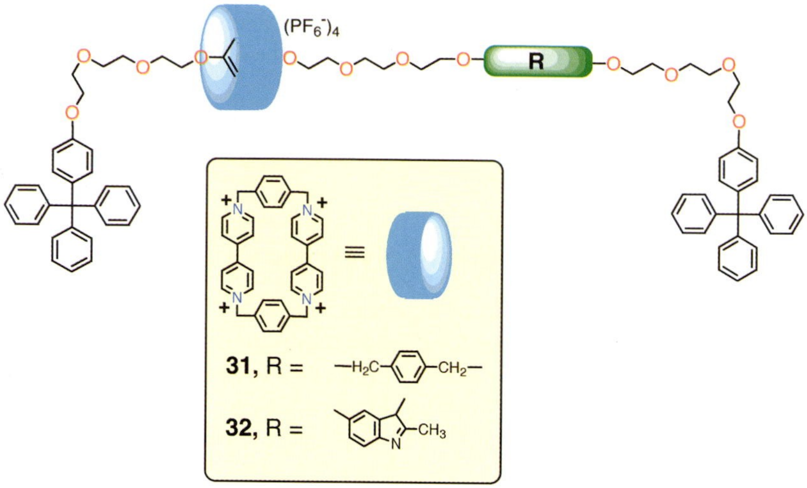

Figure 4.15 *Two 'non-symmetrical' rotaxanes*[55]

reaction would result in a shift of the system's charge-transfer band to a lower wavelength. Such a shift would provide a useful means of monitoring the behaviour of the system.

Accordingly, the required new [2]-rotaxane **32** was successfully synthesised (Figure 4.15).[55] This time, the system did indeed have the cyclophane located exclusively at one site at $-40\,^{\circ}\mathrm{C}$ – *but it was at the hydroquinone site and not the expected indole site!* It appears that steric factors win over electronic preferences in this case, although solvation effects may also play a role. Clearly, another approach was required.

Further progress was achieved with the subsequent synthesis of the bis(2-oxypropylenedithio)tetrathiafulvalene-containing [2]-rotaxane **33**.[56] An electron-rich tetrathiafulvalene derivative appeared an attractive candidate for incorporation into the linear component. The parent tetrathiafulvalene is of comparable bulk to a hydroquinone group and has also been demonstrated to exhibit reversible redox behaviour at low potentials. Further, the green 1 : 1 complex between tetrathiafulvalene and the above cyclophane tetracation had been characterised previously –

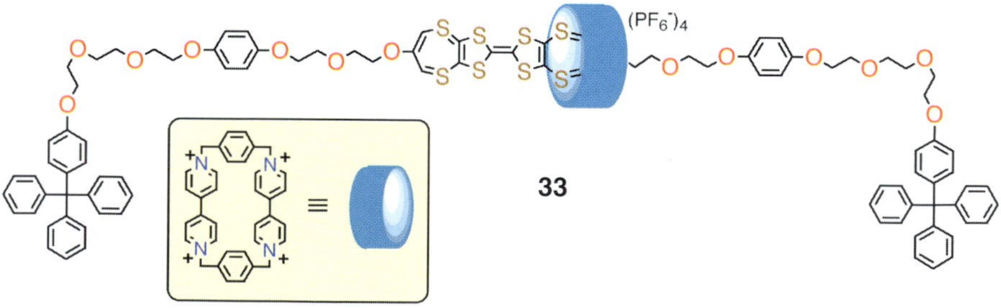

[56]   P.R. Ashton, R.A. Bissell, N. Spencer, J.F. Stoddart and M.S. Tolley, *Synlett*, 1992, 923.

as expected, the guest was shown to be inserted through the cavity of the cyclic host.[57]

Using the general synthetic approach (involving, in this case, high pressure conditions and dimethyl sulfoxide as solvent), **33** was obtained as orange crystals. The central tetrathiafulvalene nucleus was expected to be the site preferentially occupied by the tetracationic cyclobis(paraquat-*p*-phenylene) cyclophane, since optimum π-stacking interactions between host and guest should occur in this arrangement. However, once again the ^{1}H NMR investigations were somewhat of a surprise. While in deutero-dimethyl sulfoxide the tetracationic cyclophane does predominantly occupy the expected tetrathiafulvalene site; in deutero-acetone, it is clearly located solely on a hydroquinone site, indicating an unexpected sensitivity of the system to external (solvation) perturbation.

Despite the fact that none of the three non-symmetrical [2]-rotaxanes just described fully meet the goal of having the cyclophane tetracation solely occupying the π-electron-donor site with the lowest oxidation potential, in two cases this isomer is the *preferred* one – provided that an appropriate choice of solvent is used. The results of these studies are still of considerable value in a general sense since they serve to delineate some of the subtleties inherent in attempting to use this approach to obtain a controllable shuttle.[58]

Additional experiments were aimed at investigating other π-electron-rich site types that might prove suitable for incorporation into systems capable of undergoing controlled switching.[59] Thus both benzidine **34** and 4,4′-biphenol **35** were demonstrated to form stable host–guest complexes in acetonitrile with the cyclobis(paraquat-*p*-phenylene) tetracation. Although these hosts are structurally related, they have different redox properties, with **34** being much more readily oxidised than **35**.

34   **35**

The blue complex of **34** has a formation constant of 1044 dm³ mol⁻¹ while the corresponding value for the complex of **35** is 140 dm³ mol⁻¹. In both cases the NMR spectra confirm that these species are true inclusion complexes. The complexation of **34** was also investigated by voltammetry; the half-wave potential for the single electron oxidation of benzidine in acetonitrile is shifted anodically after the addition of excess host to the guest.

The related linear derivatives **36** and **37** have also been synthesised.[59] Each of these species binds to the cyclophane tetracation with an approximately similar

57 D. Philp, A.M.Z. Slawin, N. Spencer, J.F. Stoddart and D.J. Williams, *J. Chem. Soc., Chem. Commun.*, 1991, 1584.
58 R.A. Bissell and J.F. Stoddart in *Computations for the Nano-Scale*, eds. P.E. Blöchl, A.J. Fisher and C. Joachim, Kluwer, Dordrecht, 1993, pp. 141–152.
59 E. Córdova, R.A. Bissell, N. Spencer, P.R. Ashton, J.F. Stoddart and A.E. Kaifer, *J. Org. Chem.*, 1993, **58**, 6550.

36

37

binding constant to that for the corresponding parent, **34** or **35**. End-capping of the primary hydroxyl residues in each of these pseudorotaxanes, using triisopropylsilyl groups, produces the corresponding [2]-rotaxanes as green and red solids, respectively; yields were 39 and 5%. These products represented the first rotaxanes of the present class that did not incorporate hydroquinone or 1,5-naphthoquinol derivatives in the linear thread.

Based on the background studies described above, it proved possible to synthesise the non-symmetric rotaxane **38** (Figure 4.16), incorporating both benzidine and biphenol sites, in 19% yield.[46a] This system was found to be 'switchable' between the two possible translational isomers either by pH control or by electrochemical means. In both these cases the switching depends on repulsion between the cationic bead and a positive charge on the benzidine site generated by the switching process. Although the rate of interchange of the bead between 'stations' is comparable with the NMR timescale in deutero-acetonitrile at room temperature, at –44 °C the interchange becomes slow. Once again, two distinct sets of NMR signals, corresponding to the presence of both translational isomers, were observed.

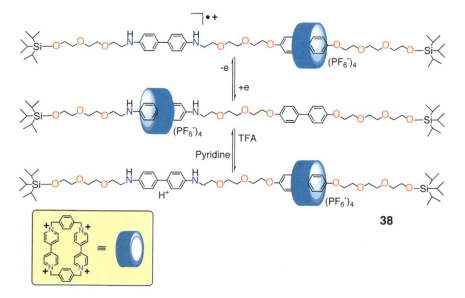

Figure 4.16 *Interconversions of the 'switchable' rotaxane* **38**[46a]

The isomers correspond to 84% occupation of the benzidine site and 16% occupation of the biphenol site. The preferential occupation of the benzidine site was expected from the prior model work, mentioned above, involving the corresponding simple benzidine and biphenol derivative guests.

In deutero-acetonitrile, protonation of the amino groups belonging to the benzidine group in **38** occurs on addition of excess deuterated trifluoroacetic acid. The resulting electrostatic barrier associated with the protonated benzidine site confines the bead to the biphenol site (Figure 4.16), resulting in a red, charge-transfer product. Neutralisation restores the original situation. It was known from the model studies that the simple benzidine-derivative charge-transfer complex is green while the corresponding biphenol complex is red.[59]

For the above mixed site system, electrochemical oxidation of the benzidine group also results in complete transfer of the cyclophane bead to the diphenol site (Figure 4.16). The initial presence of the bead on the benzidine site results in the first half-wave oxidation (to produce the corresponding cation radical) occurring at a more positive potential. In contrast, the potential of the second oxidation step (to convert the cation radical to the dication) is not affected by the presence of the charged bead since the latter is no longer associated with the site. While the properties of this system are not yet ideal for its use as a molecular switch, overall, the study represented a further step towards this goal.

In other work, the new redox-active [2]-rotaxane **39**, incorporating a 1,4-phenylenediamine subunit in its linear component, has been synthesised in high yield (Figure 4.17).[60] The redox behaviour of this product indicates considerable (increased) resistance to oxidation of the phenylenediamine subunit. This subunit appears even more promising with respect to its ability to repel the cationic cyclophane upon oxidation.

Benniston and Harriman[61] synthesised the ferrocenyl-stopped [2]-rotaxane **40** (Figure 4.18) using the Stoddart (ring-closing) procedure in order to probe its application as a model for a light-induced molecular shuttle. Previous work had demonstrated that the illumination of a tight charge-transfer complex of the type present in the above rotaxane will result in the instantaneous formation of an intimate radical ion pair (RIP).[62] In such a situation, charge recombination to form the ground state will compete with dissociation of the RIP into separated radical ions as well as with solvent infiltration to yield a solvent-separated 'loose' RIP. Use of a short laser pulse for illumination enables factors affecting the rates of charge recombination and dissociation of the RIP to be probed.

Rotaxane **40** (Figure 4.18) was synthesised since it was reasoned that it might be possible to prolong the lifetime of any redox intermediate by the incorporation of terminal redox-active stoppers. The presence of the latter may provide a means of facilitating spatial separation of the primary redox changes. Preliminary cyclic voltammetry studies indicated that the π-radical dialkoxybenzene cation, formed by excitation of the charge-transfer complex, should favour oxidation of one of the

---

60   E Córdova, R.A. Bissell and A.E. Kaifer, *J. Org. Chem.*, 1995, **60**, 1033.
61   A.C. Benniston and A. Harriman, *Angew. Chem., Int. Ed. Engl.*, 1993, **32**, 1459 and references therein.
62   T. Yabe and J.K. Kochi, *J. Am. Chem. Soc.*, 1992, **114**, 4491.

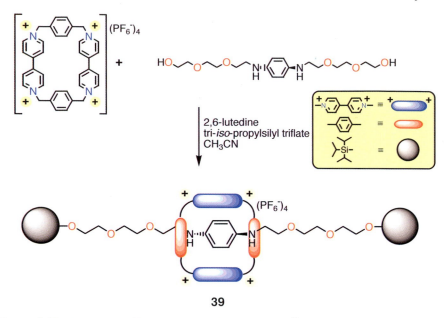

Figure 4.17 *Formation of* **39** *by threading then stoppering*[60]

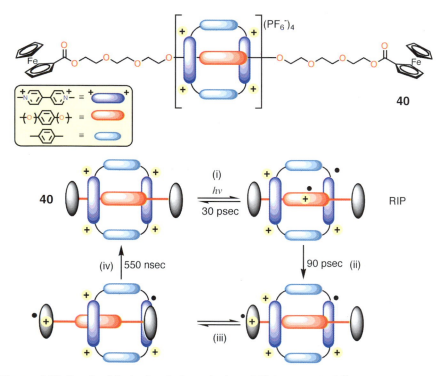

Figure 4.18 *Result of flash photolysis excitation of* **40** *in acetonitrile*[61]

appended ferrocene subunits. Flash photolysis excitation (at the wavelength of the charge-transfer absorption) of an acetonitrile solution of **40** resulted in the instantaneous appearance of a new species which, from its transient differential absorption spectrum, was assigned as the RIP formed by electron-transfer from the included dialkoxybenzene to the surrounding cyclophane. The lifetime of this excited species (22 ps) was found to be longer than that of the corresponding related [2]-catenane (9 ps) in which photo-inactive triisopropylsilyl groups replaced the terminal ferrocene units. Further, the spectrum of the product did not decay completely but yielded a residual signal, which then decayed (by first order kinetics) with a long lifetime of 550 ns.

The above observations were interpreted in the following manner (Figure 4.18). Excitation of **40** generates the RIP. The RIP is then rapidly deactivated, largely by charge recombination, but this also occurs in competition with the oxidation of one of the appended ferrocene units by the π-radical dialkoxybenzene cation generated initially. The ground state of the system is then regenerated by slow electron-transfer from the π-radical cation of the cyclic component to the adjacent ferrocenium cation.

The above process enables approximately 25% of the original redox charge to last for around 500 ns; indeed, annihilation of the remote redox charges occurs approximately 2×10^4 times slower than the primary charge recombination process. The slowness of this former process appears sufficient to allow migration of the cyclophane along the linear-component, driven by the charge repulsion between the positively charged linear and cyclic components. Such a mechanical process approximates to **40** acting as a light-induced molecular shuttle. Namely, light triggers migration of the cyclic component along the thread while the subsequent step, reverse electron-transfer, leads to its restoration to a symmetrical location with respect to the thread.

In other studies, the above research group synthesised a number of related one- and two-station, photoactive [2]-rotaxanes incorporating either ferrocene or anthracene stoppers or a combination of each type.[63] In these systems, the terminal stoppers associate with the exterior of the bis(4-4'-bipyridinium)cyclophane tetracation *via* π-stacking. Such proximity has obvious implications for the required rapid electron-transfer to a terminal (ferrocenyl) group in order to be competitive with the simple charge recombination process discussed previously. The relationship between the photoactivity, electron-transfer behaviour and concomitant configurational charges in the above systems has now been described in some detail.[64]

4.7 Use of Dialkylammonium Groups for Threading Crowns

There are a number of examples in which dialkylammonium cations have been shown to form complexes with crowns in which the NH_2^+ group is held in the plane

---

[63] A.C. Benniston, A. Harriman and V.M. Lynch, *J. Am. Chem. Soc.*, 1995, **117**, 5275 and references therein.

[64] A.C. Benniston, A. Harriman and V.M. Lynch, *J. Am. Chem. Soc.*, 1995, **117**, 5291 and references therein.

of the macrocycle such that the alkyl groups protrude on either side of the ring.[65–68] Thus, both the dibenzylammonium and di-*n*-butylammonium cations form 1 : 1 complexes with dibenzo-24-crown-8 in chloroform and dichloromethane. Complexation is characterised by the solubilisation of the hexafluorophosphate salts of these cations in the above solvents.[65] The solid state structures of these adducts confirm that the NH_2^+ centre, together with one of the adjacent CH_2 groups, lie approximately in the plane on the respective polyether macrocycles. Three point hydrogen bond formation is present: individual hydrogen bonds occur between particular ether oxygen atoms and each amine hydrogen atom as well as with one of the methylene hydrogen atoms. In the case of the dibenzylammonium complex, an additional π–π stacking interaction occurs between a phenyl group in the cation and a catechol group of the crown.

Overall, the structures of the above adducts suggested to Stoddart *et al.* that similar threading interactions could act as a basis for the production of a new family of interlocked molecules.[69] Indeed, the interaction of dialkylammonium salts and crown polyethers of the type just mentioned under various conditions has extended the range of such products: pseudorotaxane species with (host–guest) stoichiometries of 1 : 1, 1 : 2, 2 : 1 and 2 : 2 have all been characterised.[70,71] For example, when the dibenzylammonium ion interacts with the larger crown ring, bis-*p*-phenylene-34-crown-10, then the product contains two guest ammonium ions threaded simultaneously through the macrocyclic ring. The result is a double stranded system of type **41**.[70] Addition of the appropriate difunctional, bis-amine dication, as its hexafluorophosphate salt, to a solution of (i) dibenzo-24-crown-8 and (ii) bis-*p*-phenylene-34-crown-10 results in self-assembly in each case to yield the corresponding 2 : 1 and 2 : 2 adducts, whose structures are represented by **42** and **43**, respectively.

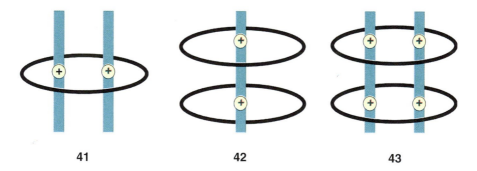

41   **42**   **43**

[65]   P.R. Ashton, P.J. Campbell, E.J.T. Chrystal, P.T. Glink, S. Menzer, D. Philp, N. Spencer, J.F. Stoddart, P.A. Tasker and D.J. Williams, *Angew. Chem., Int. Ed. Engl.*, 1995, **34**, 1865.
[66]   P.T. Glink, C. Schiavo, J.F. Stoddart and D.J. Williams, *Chem. Commun.*, 1996, 1483.
[67]   P.R. Ashton, P.T. Glink, J.F. Stoddart, S. Menzer, P.A. Tasker, A.J.P. White and D.J. Williams, *Tetrahedron Lett.*, 1996, **37**, 6217.
[68]   M.C.T. Fyfe, P.T. Glink, S. Menzer, J.F. Stoddart, A.J.P. White and D.J. Williams, *Angew. Chem., Int. Ed. Engl.*, 1997, **36**, 2068.
[69]   P.R. Ashton, E.J.T. Chrystal, P.T. Glink, S. Menzer, C. Schiavo, J.F. Stoddart, P.A. Tasker and D.J. Williams, *Angew. Chem., Int. Ed. Engl.* 1995, **34**, 1869.
[70]   P.R. Ashton, E.J.T. Chrystal, P.T. Glink, S. Menzer, C. Schiavo, N. Spencer, J.F. Stoddart, P.A. Tasker, A.J.P. White and D.J. Williams, *Chem. Eur. J.*, 1996, **2**, 709.

The X-ray structures of both these systems have been determined. In each adduct, both [N–H···O] and [C–H···O] hydrogen bonds are present between the guest(s) and oxygen atoms belonging to the encircling macrocyclic ether. For the smaller ring system, there are also π–π stacking interactions between the almost aligned catechol and *p*-xylyl units within the assembly. For both adducts, the structures in solution appear broadly similar to those observed in the solid state.

The formation of corresponding rotaxanes, both two [2]-rotaxanes and a [3]-rotaxane, *via* individual pseudorotaxane intermediates of the above type has been reported.[72] The normal strategy of 'threading then stoppering' was employed for their synthesis. However, in subsequent studies the use of a 'slippage' procedure for obtaining related products has also been described.[73] Elaborated structures related to the above but based on 'interwoven' architectures have also been synthesised.[74] An example of this type is illustrated in Figure 4.19.[75] In this, two trifurcated tris-ammonium trications of type **44** incorporating three NH_2^+-containing branches, interact with three molecules of the 34-membered crown to yield the unusual interwoven product **45**. This pseudorotaxane-related entity was isolated as its hexafluorophosphate salt.

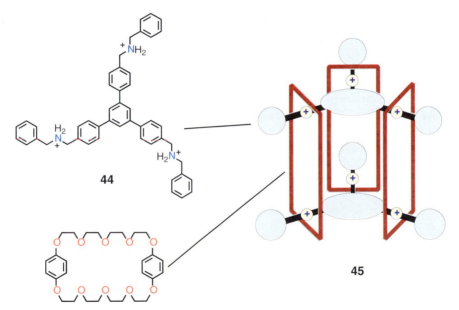

Figure 4.19 *Elaborated structure* **45** *with an interwoven architecture*[75,76]

71 P.R. Ashton, R. Ballardini, V. Balzani, M. Gomez-Lopez, S.E. Lawrence, M.V. Martinez-Diaz, M. Montalti, A. Piersanti, L. Prodi. J.F. Stoddart and D.J. Williams, *J. Am. Chem. Soc.*, 1997, **119**, 10641 and references therein.
72 P.R. Ashton, P.T. Glink, J.F. Stoddart, P.A. Tasker, A.J.P. White and D.J. Williams, *Chem. Eur. J.*, 1996, **2**, 729.
73 P.R. Ashton, I. Baxter, M.C.T. Fyfe, F.M. Raymo, N. Spencer, J.F. Stoddart, A.J.P. White and D.J. Williams, *J. Am. Chem. Soc.*, 1998, **120**, 2297.
74 P.R. Ashton, A.N. Collins, M.C.T. Fyfe, S. Menzer, J.F. Stoddart and D.J. Williams, *Angew. Chem., Int. Ed. Engl.*, 1997, **36**, 735 and references therein.
75 P.R. Ashton, A.N. Collins, M.C.T. Fyfe, P.T. Glink, S. Menzer, J.F. Stoddart and D.J. Williams, *Angew. Chem., Int. Ed. Engl.*, 1997, **36**, 59.

The self-assembly of a 'hybrid' rotaxane[76] (and related pseudocatenanes)[77] incorporating both ammonium and bipyridinium cationic units in the linear components is known. The 'switchable' [2]-rotaxane **46** incorporates one of each of these

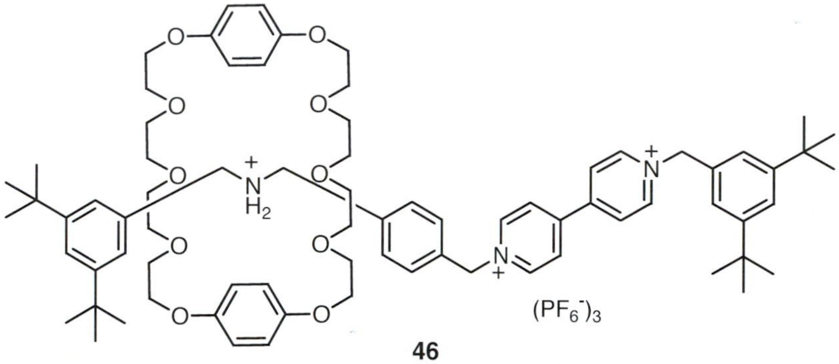

46

cationic sites in its linear strand, with the latter being encircled by a dibenzo-24-crown-8 ring.[76] An aim of this study was to develop a pH-sensitive (or alternatively, electrochemically driven) molecular switch. It was anticipated that the crown would reside initially on the NH_2^+ site but, on amine deprotonation, binding at this site would no longer be favourable and the ring would move to the bipyridinium site. This expectation was realised, with reversible pH control of the 'residence' position of the crown being cleanly achieved.

Busch *et al.*,[78] working independently of the studies just discussed, were successful in constructing the [2]-rotaxane **47** (Figure 4.20). These workers started from the required diamine-containing axle fragment and dibenzo-24-crown-8 and employed the sequence shown. The terminal amine group serves two purposes. The first is to orientate the crown correctly prior to threading and the second is to act as the site for reaction with the blocking group **48**, after threading has occurred.

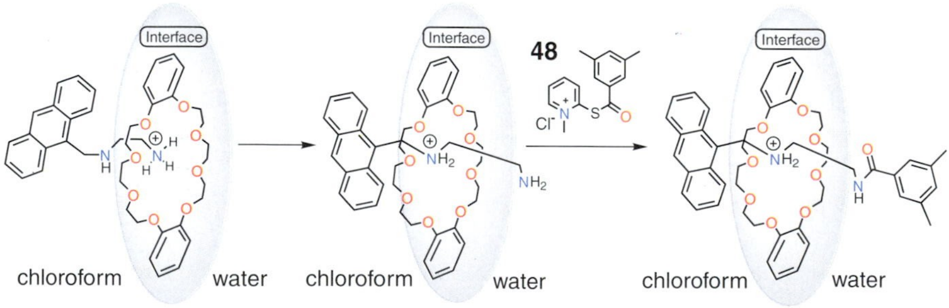

Figure 4.20 *Two-phase preparation of rotaxane* **47**[78]

[76]   M.-V. Martinez-Diaz, N. Spencer and J.F. Stoddart, *Angew. Chem., Int. Ed. Engl.*, 1997, **36**, 1904.
[77]   P.R. Ashton, P.T. Glink, M.-V. Martinez-Diaz, J.F. Stoddart, A.J.P. White and D.J. Williams, *Angew. Chem., Int. Ed. Engl.*, 1996, **35**, 1930 and references therein.
[78]   A.G. Kolchinski, D.H. Busch and N.W. Alcock, *J. Chem. Soc., Chem. Commun.*, 1995, 1289.

During the blocking step the terminal amine group is converted to a non-terminal amide linkage. The presence of the secondary amine along the chain aids the threading process by providing a site for accepting the crown after it translocates from the initial (terminal) amine site. The translocation may be driven, in part, by the possibility of π–π-stacking interactions between the crown benzo ring(s) and the anthracene terminal group at the opposite end of the axle.

A novel feature of this system is that the yield of rotaxane is increased about two-fold when a two phase (water–chloroform) solvent system is employed. The reaction proceeds on mixing the difunctional amine fragment with the crown ether in the presence of the (water-soluble) blocking reagent **48**. This acylating reagent is largely partitioned in the aqueous phase but presumably reacts at the organic–aqueous interface with the prior-threaded linear component present in the organic phase. It seems likely that this latter (threaded) component will be oriented such that its hydrophilic primary amine group is located at the biphasic interface – thus assisting easy condensation with **48**. The X-ray structure of the crystalline rotaxane confirms the predicted structure. The presence of hydrogen bonds between the secondary amine protons and ether oxygens of the crown are clearly observed, as is a contact between the anthracene and one of the crown benzo groups.

Further examples of 'hybrid' pseudorotaxanes containing a linear component incorporating two dialkylammonium centres and a 4,4'-bipyridinium unit associated with an assortment of macrocyclic polyethers have been reported. Both multi-component rotaxanes and polyrotaxanes based on the above tetracationic thread were observed to form. The reversible acid–base-controlled dethreading/rethreading of such [n]-rotaxanes has been demonstrated using a combination of NMR, absorption and fluorescence spectroscopies.[79]

4.8 Uncharged Amide-containing Rotaxanes

Although emphasis has been given so far to the self-assembly of positively charged rotaxanes, it needs to be noted that a number of uncharged rotaxanes, in which both the linear and cyclic components are neutral entities,[80] have also been synthesised by a number of workers.[81–88] Typically, these systems incorporate amide linkages and,

[79]   P.R. Ashton, R. Ballardini, V. Balzani, M.C.T. Fyfe, M.T. Gandolfi, M.V. Martinez-Diaz, M. Morosini, C. Schiavo, K. Shibata, J.F. Stoddart, A.J.P. White and D.J. Williams, *Chem. Euro. J.*, 1998, **4**, 2332.
[80]   L.F. Lindoy, *Nature*, 1995, **376**, 293.
[81]   C.A. Hunter, *J. Am. Chem. Soc.*, 1992, **114**, 5303.
[82]   S. Ottens-Hildebrandt, M. Nieger, K. Rissanen, J. Rouvinen, S. Meier, G. Harder and F. Vögtle, *J. Chem. Soc., Chem. Commun.*, 1995, 777.
[83]   A.G. Johnston, D.A. Leigh, R.J. Pritchard and M.D. Deegan, *Angew. Chem., Int. Ed. Engl.*, 1995, **34**, 1209.
[84]   F. Vögtle, M. Händel, S. Ottens-Hildebrandt, F. Ott and T. Schmidt, *Liebigs Ann.*, 1995, 739.
[85]   F. Vögtle, T. Dunnwald, M. Handel, R. Jager, S. Meier and G.Harder, *Chem. Eur. J.*, 1996, **2**, 640; A.G. Johnston, D.A. Leigh, A. Murphy, J.P. Smart and M.D. Deegan, *J. Am. Chem. Soc.*, 1996, **118**, 10662; R. Jager, S. Baumann, M. Fischer, O. Safarowsky, M. Nieger and F. Vögtle, *Liebigs Ann.*, 1997, 2269; T. Dunnwald, R. Jager and F. Vögtle, *Chem. Eur. J.*, 1997, **3**, 2043.
[86]   D.A. Leigh, A. Murphy, J.P. Smart and A.M.Z. Slawin, *Angew. Chem., Int. Ed. Engl.*, 1997, **36**, 728.

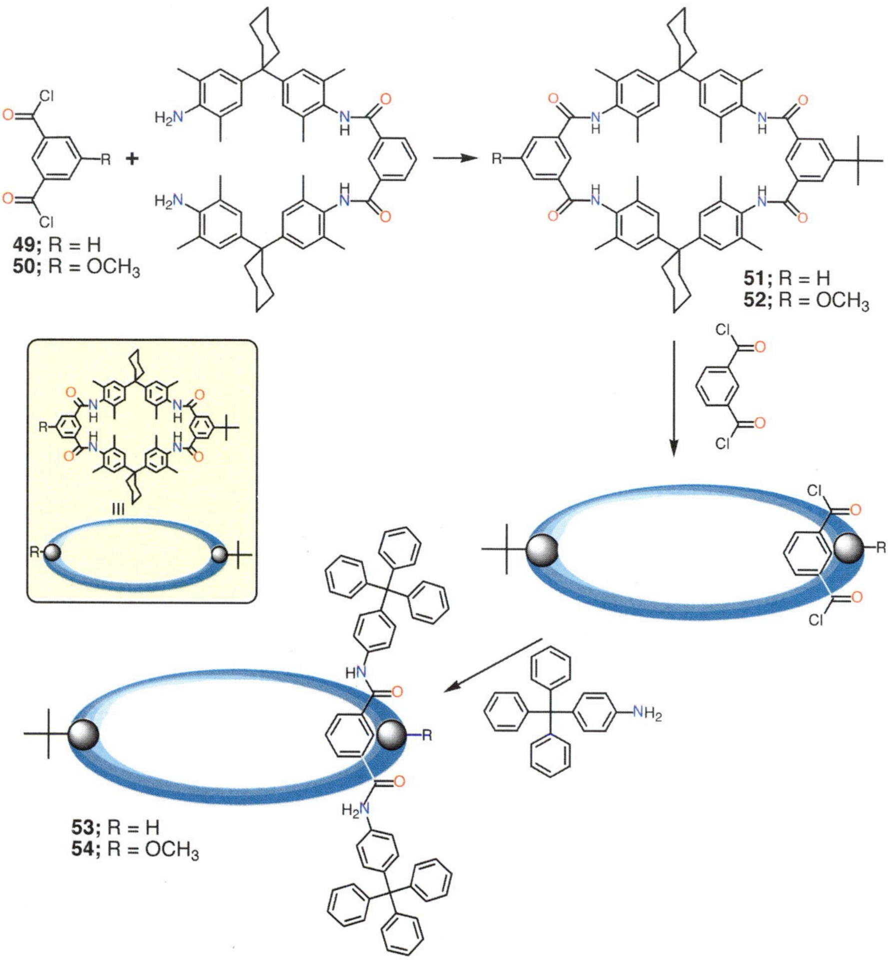

Figure 4.21 *Formation of rotaxanes* **53** *and* **54**. *Each preparation is directed by the formation of the corresponding intermediate host–guest complex shown*[84]

once again, each of the synthetic procedures generally involved a substantial element of mutual organisation of the reagents prior to formation of the interlocked products. Examples of this category are illustrated by structures **53** and **54** in Figure 4.21.[84]

The synthetic procedure outlined in the figure involves the interaction of the cyclic tetralactam species **51** and **52** with the isophthaloylic groups **49** and **50** to yield 'tight' host–guest complexes in which the guest is orientated at right angles to the plane of the ring. It appears that both weak π–π interactions and NH···O=C hydrogen bonds are again active in binding host to guest. Each end of the 'axle' was then capped using a bulky amine group. Clearly, the synthesis once again proceeds by a combination of molecular template and self-assembly processes since reaction of macrocycle **51**, the diacetyl chloride and capping amine group simultaneously in the ratio of 1 : 1 : 2 also yields **53** in 11% yield.

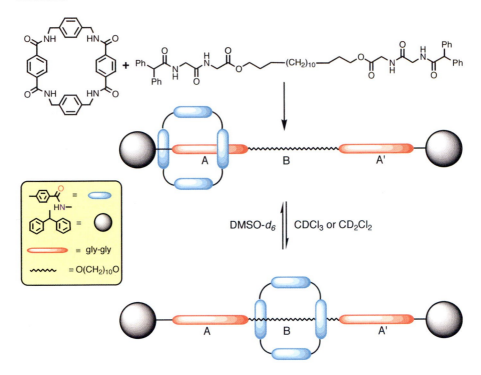

Figure 4.22 *A peptide-based molecular shuttle*[87]

Peptide-based molecular shuttles based on uncharged components have been prepared by the Leigh group[87] (Figure 4.22). NMR studies demonstrated solvent-dependent (deutero-DMSO *versus* CDCl$_3$ or CD$_2$Cl$_2$) translational isomerism of the type illustrated in the figure. While the macrocyclic component shuttles between the stations A and A' of the linear component in CDCl$_3$ or CD$_2$Cl$_2$, the major increase in the polarity of the environment on moving to deutero-DMSO results in a cessation of shuttling, with the macrocycle now spending nearly all of its time in the lipophilic region of the chain (designated B in Figure 4.22).

4.9 Cyclodextrin-based Systems

As mentioned in Chapter 3, cyclodextrins (CDs) are cyclic glucopyranose oligomers having at least 6 monomeric units linked by α-(1,4) linkages. The most common cyclodextrins contain 6, 7, and 8 units, respectively – designated α-, β- and γ-CD, respectively. The structure of α-CD is given by **55**. The cone-shape of these structures gives rise to well-defined hydrophobic cavities, with the inner surface defined by the oxygens of the glucopyranosidic linkages and the hydrogen atoms at the

[87] A.S. Lane, D.A. Leigh and A. Murphy, *J. Am. Chem. Soc.*, 1997, **119**, 11092.
[88] C. Fischer, M. Nieger, O. Mogck, V. Bohmer, R. Ungaro and F. Vögtle, *Eur. J. Org. Chem.*, 1998, 155.

55

3- and 5-positions. Secondary 2- and 3-hydroxyl groups are positioned around the wider opening of the cone while primary 6-hydroxyl groups surround the narrower opening. These hydroxyls are primarily responsible for the excellent water-solubilities of the cyclodextrins. The 6-unit system (α-cyclodextrin) has an internal diameter of about 4.5 Å which increases to about 7.0 Å in the 7-unit system (β-cyclodextrin), whereas in the 8-unit ring (γ-cyclodextrin) it is around 8.5 Å. Molecular orbital calculations based on X-ray data indicate that cyclodextrins possess a dipole moment, with the positive and negative poles located at the narrow and wide ends of the cone, respectively. The hydrophobic nature of the cavities is primarily responsible for the well documented ability of these natural receptors to bind a wide range of organic and other substrates in aqueous media. Because of this property, cyclodextrins have been used to preorganise suitable groups for the assembly of a range of rotaxanes.

Since a very large number of rotaxanes and polyrotaxanes that contain cyclodextrins and cyclodextrin derivatives are known, discussion of only a representative selection of such compounds is presented in this section.

In an early study, Ogino[89,90] was successful in preparing the first examples of cyclodextrin-containing rotaxanes; these incorporated metal complexes as terminal groups. These assemblies were obtained in 2–19% yield by threading an α- or β-cyclodextrin on to 1,10-decanediamine, 1,12-dodecanediamine or 1,14-tetradecanediamine. The respective linear chains were terminated by chlorobis(ethane-1,2-diamine)cobalt(III) species using the sequence shown in Figure 4.23. In this case the diameter of the cobalt(III) complex is larger than the cyclodextrin cavity and hence is big enough to prevent unthreading. Further, the choice of the kinetically inert (towards ligand exchange) cobalt(III) ion ensures that the 'stoppers' remained intact.

In a related study, α- or β-cyclodextrin rotaxanes based on $[(en)_2CoNH_2(CH_2)_2 S(CH_2)_n S(CH_2)_2 NH_2 Co(en)_2]Cl_6$ (where en = ethane-1,2-diamine) were synthe-

[89]   H. Ogino, *J. Am. Chem. Soc.*, 1981, **103**, 1303.
[90]   H. Ogino and K. Ohata, *Inorg. Chem.*, 1984, **23**, 3312.

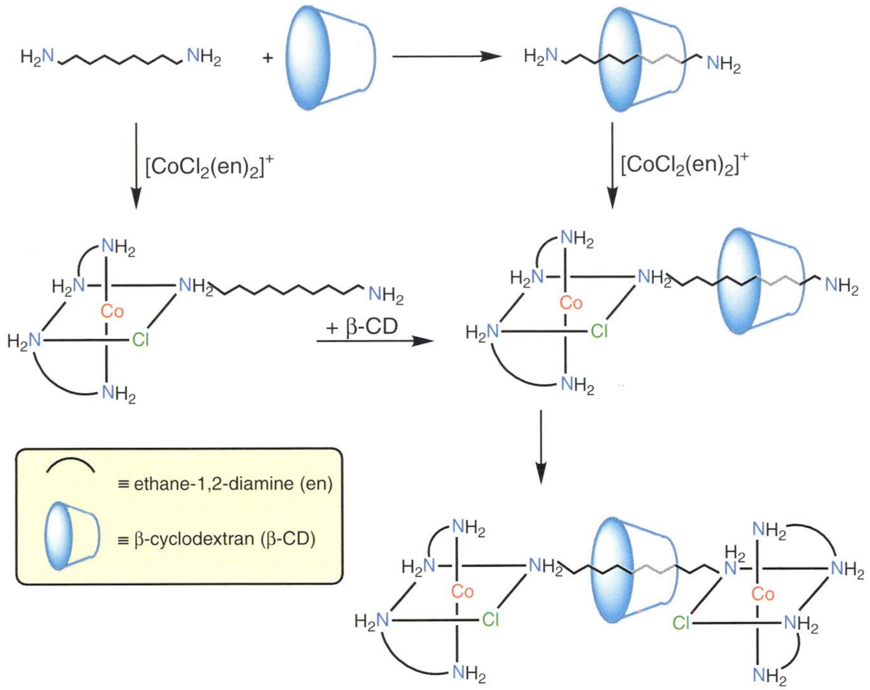

Figure 4.23 *A cyclodextran-based rotaxane stoppered with cobalt(III) complexes*[89,90]

sised.[91] Derivatives incorporating both one and two threaded cyclodextrins were obtained. Interestingly, these syntheses proceed with partial stereoselectivity with respect to the configurations of the terminal cobalt(III) species. Subsequently, other related systems were reported[92] including the non-symmetrical rotaxane derived from $[(en)_2Co\{O_2CCH_2S(CH_2)_{10}SCH_2CH_2NH_2\}Co(en)_2]^{5+}$ and α-cyclodextrin.

Since these studies, many other rotaxanes incorporating cyclodextrins or their substituted derivatives have been reported,[93] including other metal-containing systems. The latter include a series of α-cyclodextrin rotaxanes formed by the reaction of labile $[Fe(CN)_5OH_2]^{3-}$ ions with pre–threaded 1,1″-(α,ω-alkanediyl)-bis(4,4′-bipyridinium) dicationic moieties (incorporating methylene chains between 8- and 12-members long).[94]

Related symmetrical systems, involving α-cyclodextrin threaded onto alkyl chains that are terminated by cationic aromatic (heterocyclic) end groups, are also known.[95] Formation of the above species involves threading of the 4,4′-bipyridinium groups through the α-cyclodextrin such that the alkyl chain lies in the cavity, with the cationic bipyridinium end groups extending on either side.

91   K. Yamanari and Y. Shimura, *Bull, Chem. Soc. Jpn.*, 1983, **56**, 2283.
92   K. Yamanari and Y. Shimura, *Bull. Chem. Soc., Jpn.*, 1984, **57**, 1596.
93   H. Ogino, *New J. Chem.*, 1993, **17**, 683.
94   R.S. Wylie and D.H. Macartney, *J. Am. Chem. Soc.*, 1992, **114**, 3136.
95   H. Yonemura, H. Saito, S. Matsushima, H. Nakamura and T. Matsuo, *Tetrahedron Lett.*, 1989, **30**, 3143; H. Saito, H. Yonemura, H. Nakamura and T. Matsuo, *Chem. Lett.,* 1990, 535.

The pentacyanoferrate(II) group is known to have a strong affinity for aromatic *N*-heterocycles and rapidly binds to both ends of the self-assembled α-cyclodextrin/ligand complex to yield the rotaxane **56**. Alternatively, the product

56

can be self-assembled by reaction of α-cyclodextrin with a solution of the pre-formed, bridged dimeric complex – a reflection of the inherent lability of the iron(II) co-ordination spheres in the blocking end groups. Indeed, the self-assembly of **56** was shown to occur irrespective of the order of addition of the α-cyclodextrin, the linear component, and the iron(II) complex. Other related rotaxane species in which pyrazine has replaced one or both of the 4,4'-bipyridine moieties of **56** have also been synthesised.[96] The kinetics of self-assembly of these rotaxanes upon mixing the linear (dimetal) complex fragment with α-cyclodextrin is in accord with the initial rate-determining dissociation of one iron(III) complex unit, followed by cyclodextrin threading and, finally, rapid recomplexation of the 'missing' $[Fe(CN)_5]^{2-}$ group.

The rotaxane **57** has been obtained by initial formation of the 4,4'-diaminostilbene inclusion complex of β-cyclodextrin followed by reaction with the appropriate bulky blocking moieties.[97] For this system, the UV–Vis and induced circular dichroism (CD) spectra confirmed that the central aromatic chromophore of the linear component was embedded in the cyclodextrin cavity.

57

A number of other 'simple' rotaxanes based on cyclodextrins have been reported.[98–100] Rotaxanes incorporating more than one threaded cyclodextrin moiety are also known. One such system involves a linear component consisting of a 16-methylene chain unsymmetrically terminated with carbaxole and viologen

96   D.H. Macartney and C.A. Waddling, *Inorg. Chem.*, 1994, **33**, 5912.
97   M. Kunitake, K. Kotoo, O. Manabe, T. Muramatsu and N. Nakashima, *Chem. Lett.*, 1993, 1033.
98   W. Herrmann, B. Keller and G. Wenz, *Macromolecules,* 1997, **30**, 4966.
99   S. Anderson, T.D.W. Claridge and H.L. Anderson, *Angew. Chem., Int. Ed. Engl.*, 1997, **36**, 1310.
100   X.H. Shen, M. Belletete and G. Durocher, *J. Phys. Chem. B*, 1998, **102**, 1877.

groups. NMR evidence shows that two α-cyclodextrins are threaded stepwise (as well as stereoselectively) onto the linked compound.[101]

While lying outside the scope of the present discussion, it is noted that a range of polymeric systems incorporating cyclodextrin-based rotaxanes, often with unusual architectures, have been synthesised.[102]

A further type of rotaxane is represented by the threaded heptakis(2,6-*O*-methyl)-β-cyclodextrin derivative shown in Figure 4.24.[103] The diammonium dichloride salt **58** is only sparingly soluble in water but when a 1.5 molar equivalent of heptakis(2,6-*O*-methyl)-β-cyclodextrin was added, the salt dissolved fully – strongly suggesting the formation of an inclusion complex. The inclusive nature of the resulting complex **59** was confirmed using the NMR spectral changes observed for the host signals; these indicated that the cyclodextrin cavity now contained the electron-rich aryl rings belonging to the linear fragment. Addition of excess sodium tetraphenylborate to the above aqueous solution resulted in immediate precipitation of the di-tetraphenylborate salt derived from **59**. Solution studies in acetone suggest that the presence of terminal ammonium-tetraphenylborate ion pairs serve to inhibit the unthreading of the system in this solvent. The product could thus be assembled even though the respective subunits are held together solely by non-covalent forces. Further, it was isolated in high yield (71%).

Other charged rotaxanes, each incorporating an electroactive group, were derived from threading acid-derivatised alkyldimethyl(ferrocenylmethyl)ammonium cations

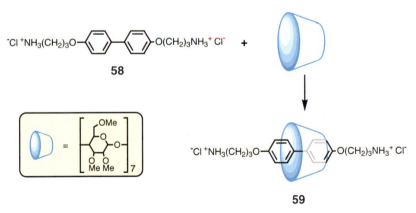

Figure 4.24 *Preparation of rotaxane* **59**[103]

[101] H. Yonemura, T. Nojiri and T. Matsuo, *Chem. Lett.*, 1994, 2097.
[102] M. Born and H. Ritter, *Makromol. Chem., Rapid Commun.*, 1991, **12**, 471; H. Ritter, *Macromol. Symp.*, 1994, **77**, 73; M. Born and H. Ritter, *Angew. Chem., Int. Ed. Engl.*, 1995, **34**, 309; I. Yamaguchi, K. Osakada and T. Yamamoto, *J. Am. Chem. Soc.*, 1996, **118**, 1811; A. Harada, M. Kamachi, *Macromolecules*, 1990, **23**, 2821; A. Harada, J. Li, M. Kamachi, *Nature*, 1992, **356**, 325; A. Harada, J. Li, T. Nakamitsu and M. Kamachi, *J. Org. Chem.*, 1993, **58**, 7524; A. Harada, J. Li and M. Kamachi, *Nature*, 1993, **364**, 516; A. Harada, J. Li, S. Suzuki and M. Kamachi, *Macromolecules*, 1993, **26**, 5267; A. Harada and M. Kamachi, *J. Chem. Soc., Chem. Commun.*, 1990, 1322; G. Wenz and B. Keller, *Angew. Chem., Int. Ed. Engl.*, 1992, **31** 197.
[103] T. Venkata, S. Rao and D.S. Lawrence, *J. Am. Chem. Soc.*, 1990, **112**, 3614.

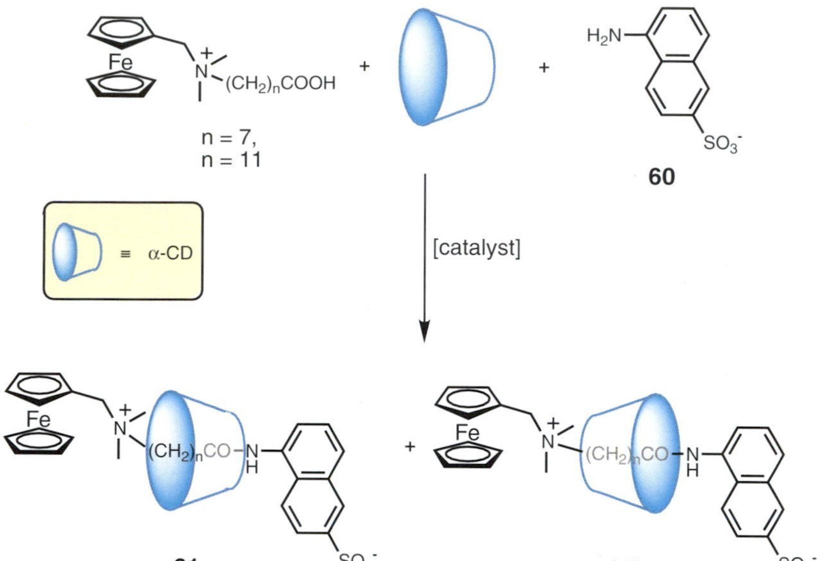

Figure 4.25 *Formation of the isomeric redox-active rotaxanes* **61a** *and* **61b**[104]

of different chain lengths through α-cyclodextrin,[104] followed by condensation of the terminal carboxylic acid groups with the amine group of the naphthalene-2-sulfonate derivative **60** (Figure 4.25). The reaction was performed in the presence of 1-[3-(dimethylamino)propyl]-3-ethylcarbodiimide hydrochloride as catalyst. The products were two sets of isomeric rotaxanes of form **61a** and **61b**. Interestingly, the larger β-cyclodextrin shows different behaviour: it does not 'thread' onto the alkyl chain but interacts preferentially with the ferrocene subunit. There is strong NMR evidence that each of the product zwitterionic asymmetric rotaxanes exists as a pair of positional isomers, reflecting different orientations of the cyclodextrin cone on the respective linear components.

The lipophilic cyclodextrin **62** has been demonstrated to act as a host molecule for the charged guests of type **63** and **64** in acetonitrile (Figure 4.26).[105] The binding constants are 150 and 450 dm^3 mol^{-1}, respectively. Attempts to quaternise the pyridine unit of **63** using **65** in the presence of bound **62** gave, however, the corresponding unthreaded product. In contrast, with **64**, the required rotaxane **66** was obtained in 20% yield. Clearly, the length of the spacer group is critical to the success of the synthesis in this case. In an extension of this study, Wenz *et al.*[106] were successful in devising a related (improved) synthesis which led to the corresponding [2]-rotaxane in which the ester groups of the linear component in **66** were

104 R. Isnin and A.E. Kaifer, *J. Am. Chem. Soc.*, 1991, **113**, 8188; R. Isnin and A.E. Kaifer, *Pure Appl. Chem.*, 1993, **65**, 495.
105 G. Wenz, E. von der Bey and L. Schmidt, *Angew. Chem., Int. Ed. Engl.*, 1992, **31**, 783.
106 G. Wenz, F. Wolf, M. Wagner and S. Kubik, *New J. Chem.*, 1993, **17**, 729.

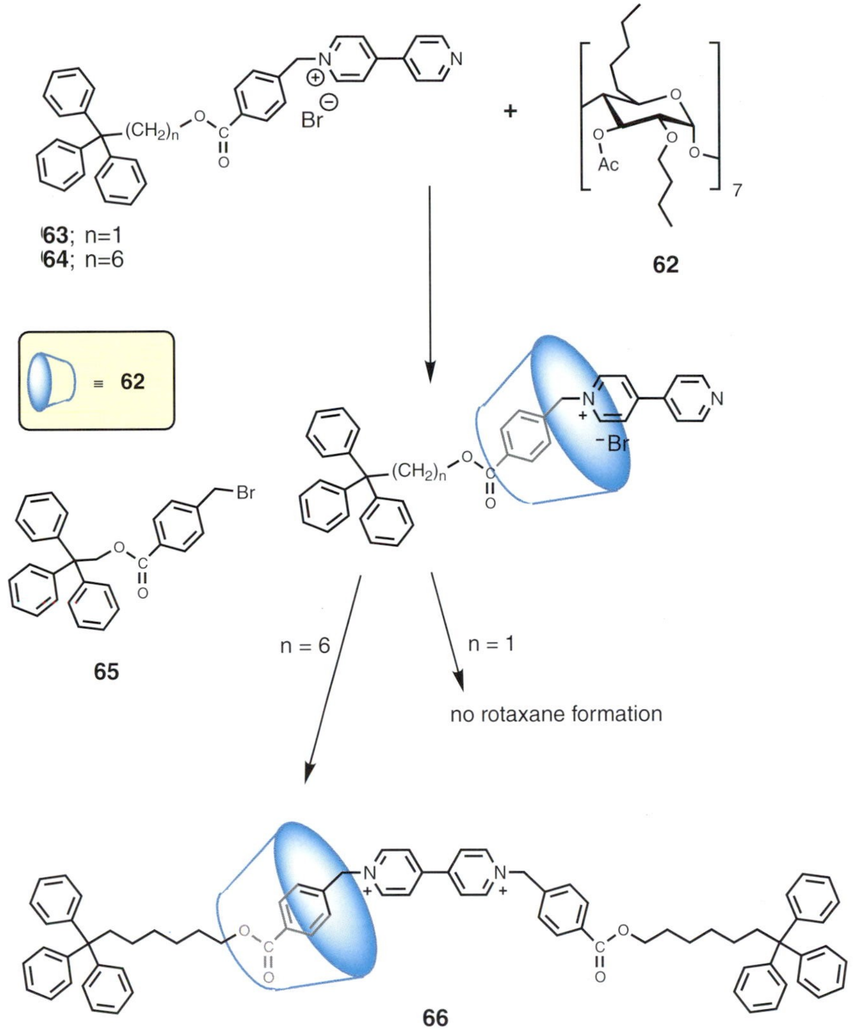

Figure 4.26 *Formation of the cationic* [2]-*rotaxane* **66**[105]

replaced by ether groups. In this system, both cationic centres on the linear component may act as specific binding sites for the acetyl groups appended to the periphery of the cyclodextrin cone. Physical measurements indicate that the threaded cyclodextrin prefers to be positioned off-centre towards one side of the linear component.

Other related studies concerned with the effects of cyclodextrin complexation on the properties of surfactants of various *n*-alkyl chain lengths, also incorporating a covalently attached electroactive group, have been reported.[107] Thus, the nature of

[107] R. Isnin, C. Salam and A.E. Kaifer, *J. Org. Chem.*, 1991, **56**, 35 and references therein.

a range of host–guest adducts between α-, β-, and γ-cyclodextrins and alkyl-dimethyl(ferrocenylmethyl)ammonium salts has been probed using both electrochemical and NMR methods. The alternate modes of binding of cyclodextrins of different ring size were employed to construct small supramolecular assemblies. For example, this is exemplified by the isolation of a quaternary complex between a linear C_{16}-alkyl derivative of ferrocene and two α-cyclodextrin molecules as well as a β-cyclodextrin molecule. In this species, the two α-cyclodextrin molecules appear to thread onto the hexadecyl aliphatic chain while the β-cyclodextrin simultaneously encloses the ferrocene subunit.

Finally, as mentioned earlier, a considerable number of other cyclodextrin-based rotaxane and pseudorotaxane systems have been reported[96,108] but, by and large, they represent 'variations on a theme' with respect to the individual systems discussed in this section.

[108] See for example: M. Kunitake, K. Kotoo, O. Manabe, T. Muramatsu and N. Nakashima, Chem. Lett., 1993, 1033; A. Toki, H. Yonemura and T. Matsuo, Bull. Chem. Soc. Jpn., 1993, 66, 3382; H. Yonemura, T. Nojiri and T. Matsuo, Chem. Lett., 1994, 2097; A. Harada, J. Li and M. Kamachi, Nature, 1994, 370, 126; M. Born, T. Koch and H. Ritter, Acta Polym., 1994, 45, 68; D. Armspach, P.R. Ashton, R. Ballardini, V. Balzani, A. Godi, C.P. Moore, L. Prodi, N. Spencer, J.F. Stoddart, M.S. Tolley, T.J. Wear and D.J. Williams, Chem. Eur. J., 1995, 1, 33; M. Born, T. Koch and H. Ritter, Macromol. Chem. Phys., 1995, 196, 1761; M. Born and H. Ritter, Adv. Mater., 1996, 8, 149; T. Kuwabara, A. Nakamura, T. Ikeda, H. Ikeda, A. Ueno and F. Toda, Supramolecular Chem., 1996, 7, 235; D.H. Macartney, J. Chem. Soc. Perkin Trans. 2, 1996, 2775; A.P. Lyon and D.H. Macartney, Inorg. Chem., 1997, 36, 729; G. Pistolis and A. Malliaris, J. Phys. Chem. B, 1998, 102, 1095; A.P. Lyon, N.J. Banton and D.H. Macartney, Can. J. Chem., 1998, 76, 843.

CHAPTER 5

Catenanes

5.1 Introduction

The early synthesis of rotaxane species of the type discussed in Chapter 4 provided the inspiration for extending the work towards more ambitious goals and led directly to what Fraser Stoddart subsequently referred to as 'looping the loop'. Namely, the formation of a catenane.[1]

Catenanes (Latin: *catena*, chain) may be defined as linked supramolecular systems consisting of cyclic subunits, not necessarily identical, held together by mechanical means in an analogous fashion to the links in a chain. Usually the components also show considerable mutual association *via* one or more non-covalent interactions.

At present, the known catenanes can be divided into two categories – those prepared by metal template synthesis and those synthesised in the absence of a metal-ion influence. A considerable number of catenanes of the first type have been prepared by Sauvage *et al.* as well as by a number of other workers. However, discussion on these important metal-ion-directed systems is deferred to the next chapter in which particular supramolecular assemblies produced by metal-ion-controlled procedures are discussed.

5.2 Statistical Threading

The concept of producing molecular systems linked in the above mechanical manner is not new.[2] In fact, the construction of a catenane has been pre-empted by Nature. The occurrence of duplex circular DNA in a catenane form is well established – with the interlinking of two double stranded DNA rings being effected enzymatically.[3]

The initial synthesis of a [2]-catenane (that is, a system involving two interlinked rings) produced a product in which both rings were 34-membered (Figure 5.1).[4] The procedure[5] was based on the threading, controlled by statistics, of a non-cyclic

[1]   J.F. Stoddart, *Chem. Br.*, 1991, 714.
[2]   G. Schill, *Catenanes, Rotaxanes and Knots,* Academic Press, New York, 1971.
[3]   W.M. Stark, C.N. Parker, S.E. Halford and M.R. Boocock, *Nature*, 1994, **368**, 76.
[4]   E. Wasserman, *J. Am. Chem. Soc.*, 1960, **82**, 4433.
[5]   H.L. Frisch and E. Wasserman, *J. Am. Chem. Soc.*, 1961, **83**, 3789; E. Wasserman, *Sci. Am.* 1962, **207**, 94.

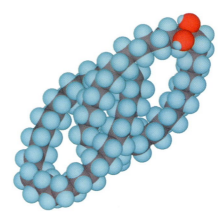

Figure 5.1 *The [2]-catenane containing linked 34-membered ring synthesised by Wasserman.[5] This was the first characterised catenane; it was prepared by statistical threading, not self-assembly*

component into a second cyclic species, followed by ring closure. Largely reflecting the random nature of the threading process, the yield of required product was extremely low. Subsequent studies, using a technique involving the macrocycle immobilised on a polymer, gave a higher yield (but still only 7%) on release of the product from the polymer.[6] Further development of the statistical threading approach subsequently resulted in yields of about 10–14%.[7,8] In other experiments, the statistical threading of polyethylene glycols through dibenzocrown polyethers has been demonstrated,[7,9] with the first hydrocarbon catenane also being prepared by a similar procedure.[10]

Other procedures – often multistep – have also been employed to obtain catenanes.[11] However, although useful as a comparison, these procedures are of marginal relevance to the present discussion since they are not primarily based on self-assembly principles.[12]

5.3   Charged Catenanes

In 1989, following extensive preparatory work discussed, the template-directed synthesis of the positively charged [2]-catenane **5** was reported by the Stoddart group (Figure 5.2).[13] Only mild conditions were required, with the crystalline catenane

6   I.T. Harrison and S. Harrison, *J. Am. Chem. Soc.*, 1967, **89**, 5723.
7   G. Agam and A. Zilkha, *J. Am. Chem. Soc.*, 1976, **98**, 5214.
8   G. Schill, W. Beckmann, N. Schweickert and H. Fritz, *Chem. Ber.*, 1986, **119**, 2647.
9   G. Agam, D. Graiver and A. Zilkha, *J. Am. Chem. Soc.*, 1976, **98**, 5206.
10 G. Schill, N. Schweickert, H. Fritz and W. Vetter, *Angew. Chem., Int. Ed. Engl.*, 1983, **22**, 889.
11 (a) G. Schill and A. Lüttringhaus, *Angew. Chem., Int. Ed. Engl.*, 1964, **3**, 546; (b) G. Schill and H. Zollenkopf, *Liebigs Ann. Chem.*, 1969, **721**, 53; (c) G. Schill and C. Zurcher, *Chem. Ber.* 1977, **110**, 2046; (d) K. Rißler, G. Schill, H. Fritz and W. Vetter, *Chem. Ber.* 1986, **119**, 1374; (e) G. Schill, N.H. Schweickert, H. Fritz and W. Vetter, *Chem. Ber.* 1988, **121**, 961.
12 M. Belohradsky, F.M. Raymo and J.F. Stoddart, *Czech. Chem. Commun.*, 1997, **62**, 527.
13 P.R. Ashton, T.T. Goodnow, A.E. Kaifer, M.V. Reddington, A.M.Z. Slawin, N. Spencer, J.F. Stoddart, C. Vicent and D.J. Williams, *Angew Chem., Int. Ed. Engl.*, 1989, **28**, 1396.

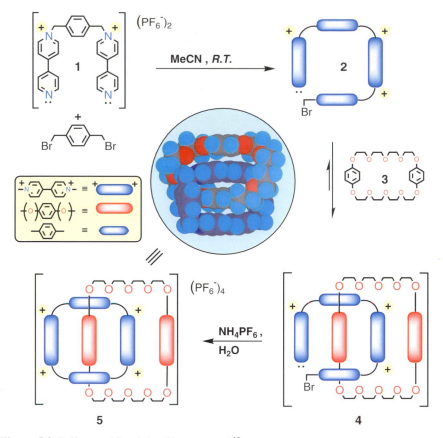

Figure 5.2 *Self-assembly of the* [2]*-catenane* **5**[13]

being obtained in 70% yield after an overnight reaction. Subsequently, it was shown that a more 'extensive' assembly process could be employed to prepare **5**. Thus, stirring a solution containing stoichiometric amounts of *p*-dibromoxylene, the macrocycle **3** and 4,4'-bipyridine in *N,N*-dimethylformamide at room temperature for five days also yielded **5**. However, after recrystallisation of the product the yield was only 18% in this case.[14]

The X-ray structure of the crystalline product (Figure 5.2) confirmed the presence of the expected alternating π–π donor–acceptor stacking. The mean separation of the two hydroquinone derivatives is 7.0 Å while the corresponding separation between the bipyridinium rings is 7.1 Å – distances which strongly suggest the presence of significant π-stacking interactions between adjacent rings.[15] There are also weak edge-to-face interactions between the *p*-phenylene units of the

14   C.L. Brown, D. Philp and J.F. Stoddart, *Synlett*, 1991, 459.
15   P.L. Anelli, P.R. Ashton, R. Ballardini, V. Balzani, M. Delgado, M.T. Gandolfi, T.T. Goodnow, A.E. Kaifer, D. Philp, M. Pietraszkiewicz, L. Prodi, M.V. Reddington, A.M.Z. Slawin, N. Spencer, J.F. Stoddart, C. Vicent and D.J. Williams, *J. Am. Chem. Soc.*, 1992, **114**, 193.

tetracations and the included hydroquinone rings. Adsorption and luminescence spectral studies clearly showed the presence of the expected charge-transfer interaction between the component parts of this product. The catenane exhibits some modification of the intense π–π^* band of the free bis(bipyridinium) tetracationic macrocycle **6**; there is also the appearance of new broad (but weak) bands in the

visible region, as well as the quenching of the luminescence of the hydroquinone units (observed for the free bis-*p*-phenylene-34-crown-10). Electrochemical and ESR evidence also confirm the presence of significant mutual electronic interaction between the interlinked rings. This interaction is composed of both electrostatic and dispersive components, with the charge-transfer interactions between the π-electron-deficient bipyridinium systems and the π-electron-rich hydroquinone rings making a significant overall contribution.

As a consequence of their different environments in the catenane, the two-bipyridinium units undergo reduction at different potentials. It was proposed that stepwise behaviour of this type may be able to be exploited in the design of future molecular (electronic) devices.

The remarkable self-assembly reaction discussed above appears to proceed by the following sequence. First, *p*-dibromoxylene condenses (presumably by S$_\text{N}$2 displacement) with one end of the available dicationic bipyridinium derivative **1** to produce the trication **2**. This then threads through the 34-membered dibenzo-crown macrocycle **3** in an analogous manner to that observed for the paraquat dication discussed in Chapter 4 (see pages 48–53). The success of this 'threading' step is, of course, of over-riding importance to the overall strategy. Not only do the π–π interactions serve to orientate the self-assembling components, π-donation from the hydroquinone groups of the crown ether may also help promote the final (ring-closing) step. The latter involves the formation of the second N–C bond to yield the tetracationic catenane **5** as its (tetra) hexafluorophosphate salt. A feature of this pathway is that covalent bonding alternates with, and is directed by, non-covalent bonding in a stepwise fashion. As just indicated, an integral part of this process is the setting up of π-interactions (which appear to involve associations of both the 'edge-to-face' as well as the more familiar stacking type) that control the relative orientations of the assembling components. The entire sequence serves as a good illustration of what can be achieved, given sufficient foresight.

Other experiments involved competitive self-assembly in which the precursors of [2]-catenanes of the above and closely related type were given some choice as to which components would ultimately be incorporated into the catenated structure.[16] This study indicated that the selectivities associated with the self-assembly process appear not to be controlled by the thermodynamics of the non-covalent interactions but rather be governed by the relative rates of covalent bond formation in the final ring-closing step(s) in forming the catenane.

Dynamic NMR spectroscopy indicated that the macrocyclic crown component in the [2]-catenane **5** is revolving through the tetracationic cyclophane ring around 300 times per second at 25 °C while it is simultaneously pirouetting around it at about 2000 times per second.[17]

In a further development, the effect of replacing the stepwise procedure discussed above by an *in situ* one involving the reaction of *p*-dibromoxylene **7**, the bis-*p*-phenylene-34-crown-10 **3**, and 4,4'-bipyridine **8** in stoichiometric quantities was investigated (Figure 5.3).[14] The reaction mixture was stirred at room temperature in *N,N*-dimethylformamide for five days. At the end of this period a deep red precipitate had formed. Treatment of this product dissolved in water with ammonium hexafluorophosphate and subsequent purification of the resulting solid yielded the [2]-catenane **5** in 18% yield. While this reaction represents a significant example of self-assembly, when the pressure was increased to 12 kbar (at 15 °C for 24 hours) then the degree of self-assembly became even more significant. The yield of purified product then rose to 42%. The latter increase may be rationalised in terms of an enhancement in the magnitude of the association constant for the key intermediate **4** (Figure 5.2) with increase in pressure. It has been well documented that higher pressures enhance quaternisation rates at nitrogen.[18]

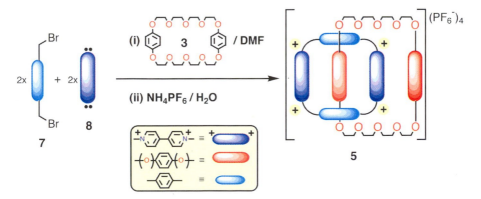

Figure 5.3 *Self-assembly by* **5** *from five components*

[16] D.B. Amabilino, P.R. Ashton, L. Perez-Garcia and J.F. Stoddart, *Angew. Chem., Int. Ed. Engl.*, 1995, **34**, 2378.
[17] J.F. Stoddart, *An. Quim.*, 1993, **89**, 51 and references therein.
[18] P. Ostazewski and M. Pietraszkiewicz, *J. Inclusion Phenom.*, 1987, **5**, 663.

Substitution of bis-1,5-naphtho-38-crown-10 for the bis-*p*-phenylene-34-crown-10 also leads to the formation of the corresponding [2]-catenane.[19,20] Again this species may be obtained by a spontaneous self-assembly process in which the dipyridinium salt **1**, *p*-dibromoxylylene and the dinaphtho crown are stirred together in acetonitrile at room temperature. Within two hours the solution turns deep purple and after two days the crude bright purple product can be isolated by removal of the solvent. On purification and addition of ammonium hexafluorophosphate, the catenane was obtained in 51% yield. The X-ray crystal structure revealed (Figure 5.4) an interlocked and, as expected, highly ordered molecular arrangement in which one 1,5-naptho residue is sandwiched between the two bipyridinium groups of the tetracationic macrocycle. The remaining naptho residue is on the outside and immediately adjacent to one of these bipyridinium groups. There is near parallel alignment of all four aromatic rings, with an overall arrangement that indicates the presence of strong π-interactions. The mean separation between aromatic rings is around 3.4–3.5 Å. In addition, there is once again clear evidence for strong aromatic–aromatic, edge-to-face π-interactions involving the 'internal' naphtho aromatic ring which has two of its hydrogens directed orthogonally into the centres of the *p*-phenylene rings belonging to the tetracationic macrocycle. The centroid-to-centroid separation between the individual protons and the aromatic rings is around 4.7 Å.

A variable temperature ¹H NMR study of the above compound has been undertaken. Despite the presence of a surfeit of exchange processes, it was concluded that this [2]-catenane is a highly ordered molecular assembly in solution.[20]

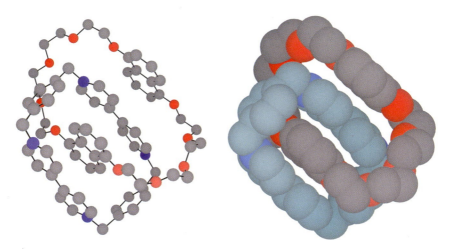

Figure 5.4 *X-Ray structure of* [2]-*catenane formed between* 1,5-*dinaphtho*-38-*crown*-10 *and* **6**; **6** *shaded in blue in the space-filling representation*[19]

19   J.Y. Ortholand, A.M.Z. Slawin, N. Spencer, J.F. Stoddart and D.J. Williams, *Angew. Chem., Int. Ed. Engl.*, 1989, **28**, 1394.
20   P.R. Ashton, C.L. Brown, E.J.T. Chrystal, T.T. Goodnow, A.E. Kaifer, K.P. Parry, D. Philp, A.M.Z. Slawin, N. Spencer, J.F. Stoddart and D.J. Williams, *J. Chem. Soc., Chem. Commun.*, 1991, 634.

Subsequently, a significant number of other catenanes have been obtained using the same general synthetic procedure – but derived from different crown ethers and the same or related tetracationic macrocycles.[21,22] The extension of the initial success to the production of these new assemblies clearly owes much to the *basic understanding* gained in the prior studies.

The prospect of using a catenane as a molecular device, such as a switch, will in the end depend upon the ability to exercise precise control over the adoption of alternative structures – such as those corresponding to translational isomers. Translational isomers may in theory occur for any [2]-catenane in which one (or both) of the component rings is unsymmetrical. There are many examples of isomerism of this type involving [2]-catenanes.[23] For example, similar template-directed procedures to those described previously have been employed to obtain the two isomeric [2]-catenanes illustrated in Figure 5.5.[24] Each isomer incorporates the same unsymmetrical polyether macrocycle, *(p*-phenylene-*m*-phenylene)-33-crown-10, interlinked with either cyclobis(paraquat-*p*-phenylene) or cyclo(paraquat-*p*-phenylene)–(paraquat-*m*-phenylene) to yield **9** and **10**, respectively. Both compounds show a strong preference to exist as the corresponding translational isomer in which the hydroquinone ring of the macrocyclic polyether, rather than its resorcinol ring, is located inside the respective cavities of the isomeric tetracationic cyclophanes. However, at low temperatures ¹H NMR evidence indicated the presence in equilibrium of a few percent of the other (less-favoured) translational isomer in which the resorcinol group occupies the inside position of the cyclophane.

Parallel studies involving the symmetrical crown, bis-*p*-phenylene-34-crown-10, and the corresponding three tetracationic cyclophanes (Figure 5.6) derived from

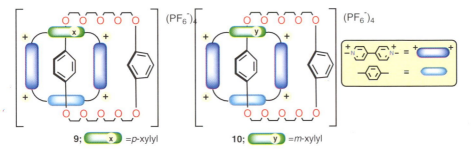

9; ⬭ x ⬭ =*p*-xylyl   **10;** ⬭ y ⬭ =*m*-xylyl

Figure 5.5 *The unsymmetric catenanes* **9** *and* **10**[24]

21   P.R. Ashton, M.A. Blower, S. Iqbal, C.H. McLean, J.F. Stoddart, M.S. Tolley and D.J. Williams, *Synlett*, 1994, 1059; P.R. Ashton, M.A. Blower, C.H. McLean, J.F. Stoddart and M.S. Tolley, *Synlett*, 1994, 1063; P.R. Ashton, J.A. Preece, J.F. Stoddart, M.S. Tolley, A.J. White and D.J. Williams, *Synthesis*, 1994, 1344.
22   D.B. Amabilino, P.R. Ashton and J.F. Stoddart, *Supramol. Chem.*, 1995, **5**, 5.
23   D.B. Amabilino, P.-L. Anelli, P.R. Ashton, G.R. Brown, E. Córdova, L.A. Godinez, W. Hayes, A.E. Kaifer, D. Philp A.M.Z. Slawin, N. Spencer, J.F. Stoddart, M.S. Tolley and D.J. Williams, *J. Am. Chem. Soc.*, 1995, **117**, 11142.
24   D.B. Amabilino, P.R. Ashton, G.R. Brown, W. Hayes, J.F. Stoddart, M.S. Tolley and D.J. Williams, *J. Chem. Soc., Chem. Commun.*, 1994, 2479.

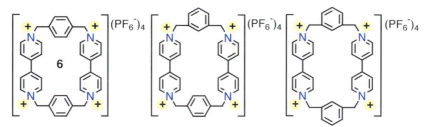

Figure 5.6 *Three tetracationic cyclophanes used to prepare catenanes with
bis-*p*-phenylene-34-crown-10*

combinations of *m*- and *p*-dibromoxylyl linker groups, have been performed.[25] From
comparative solution and solid state (X-ray diffraction) studies, it was possible to
draw the following conclusions: (i) The self-assembly process in these cases is
extremely sensitive to constitutional changes in the tetracationic cyclophane; (ii)
changing the constitution of the (π-electron-deficient) cyclophane greatly influences
the dynamic solution behaviour of the different isomeric [2]-catenanes; and (iii) the
solid state structures of these species, as well as the arrangement adopted in the
respective crystal lattices, is quite isomer-dependent. In general terms, such marked
differences in the properties of otherwise closely-related structures may possibly
be an important consideration in the design of new materials for use in molecular
devices.

   Another example of a translational isomer type is illustrated in Figure 5.7. In
this case isomerism is possible because of the unsymmetrical nature of the ethylene glycol links in the cyclic polyether.[26] At room temperature the respective cationic isomers **11** and **12** were observed to occur in the approximate ratio 64 : 36 in
both deuterated acetone and acetonitrile.

   A related [2]-catenane incorporating the unsymmetrical 1,5-naphtho-*p*-pheny-
lene-36-crown-10 and the cyclobis(paraquat-*p*-phenylene) tetracation has been

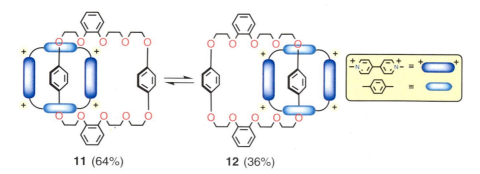

11 (64%)   **12 (36%)**

Figure 5.7 *The translational isomers **11** and **12***[26]

[25] D.B. Amabilino, P.R. Ashton, M.S. Tolley, J.F. Stoddart and D.J. Williams, *Angew. Chem., Int.
      Ed. Engl.*, 1993, **32**, 1297.
[26] D.B. Amabilino and J.F. Stoddart, *Recl. Trav. Chim. Pays-Bas.*, 1993, **112**, 429.

investigated.[27] This species also exists in solution as a mixture of translational iso-mers that depend upon whether the hydroquinone group or the 1,5-dioxynaphtha-lene group of the cyclic polyether occupies the central cavity of the tetracationic cyclophane. In this study, a strong dependence of isomer ratio on the dielectric con-stant of the solvent was observed – with the dependence ascribed to competition between the respective possible π–π interactions and the solvation of the compo-nents undergoing these interactions. That is, for this system, solvational control of isomer ratio is possible.

In an attempt to improve control over the relative amounts of translational isomers formed, systems in which the binding abilities of both the π-donor and π-acceptor units were varied have been synthesised. One example is the [2]-catenane **13** in which all four building blocks are different.[28] Since this catenane incorporates four different π-systems, four different transitional isomers are possible. However, in

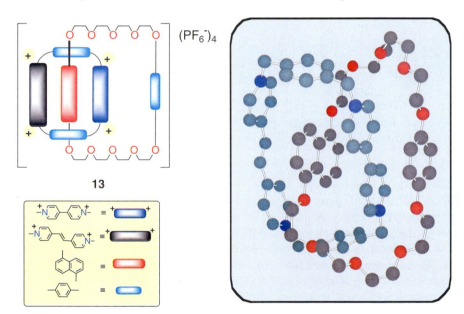

13

solution at low temperatures, only the isomer shown was observed and this is the same isomer that was found to occur in the solid by X-ray diffraction.

If molecules such as these are to be employed in molecular devices, one of the challenges is to learn how to use external stimuli to interact with the system. Such stimuli could be chemical, electrochemical, or photochemical in nature. In view of this, the two [2]-catenanes **14** and **15** were synthesised (Figure 5.8).[29] Related to their characteristic redox properties, the precursor vinylogous viologen building

[27] P.R. Ashton, M. Blower, D. Philp, N. Spencer, J.F. Stoddart, M.S. Tolley, R. Ballardini, M. Ciano, V. Balzani, M.T. Gandolfi, L. Prodi and C.H. McLean, *New J. Chem.*, 1993, **17**, 689.

[28] P.R. Ashton, L. Pérez-Garcia, J.F. Stoddart, A.J.P. White and D.J. Williams, *Angew. Chem., Int. Ed. Engl.*, 1995, **34**, 571.

[29] P.R. Ashton, R. Ballardini, V. Balzani, M.T. Gandolfi, D.J.-F. Marquis, L. Pérez-Garcia, L. Prodi, J.F. Stoddart and M. Venturi, *J. Chem. Soc., Chem. Commun.*, 1994, 177.

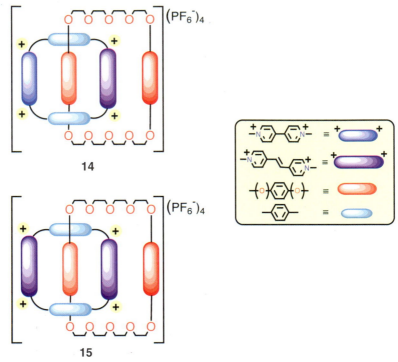

Figure 5.8 [2]-*Catenanes incorporating one and two bis(pyridinium)ethylene groups*[29]

block had previously been demonstrated to respond to stimuli of the above type. In preliminary studies, the position of equilibrium between the translational isomers of **14** was demonstrated to be quite temperature-dependent in deutero-acetone, with the isomer in which the bis-*p*-phenylene-34-crown-10 macrocycle encircles the simple bipyridinium unit being strongly preferred at low temperatures. On electrochemical reduction of **14**, it was demonstrated that initial reduction of the bipyridinium unit occurs thus promoting formation on the corresponding product in which the bis(pyridinium)ethylene unit occupies the 'inside' position. In a second redox step this latter product, by analogy with the electrochemical behaviour of **15**, then appears to be reduced. In general terms, behaviour of this type once again points the way towards the fabrication of an electrochemical 'switch'.

In an extension of the previous studies involving **13**, it has been demonstrated that a number of other [2]-catenanes can be synthesised using self-assembly procedures in which the donor macrocyclic polyethers incorporate both hydroquinone and 1,5-dioxynaphthalene units, while the acceptor tetracationic cyclophanes contain bipyridinium and/or its extended analogue, bis(pyridinium)ethylene.[30] Although the *trans* carbon–carbon double bond in the bis(pyridinium)ethylene units are

30   P.R. Ashton, R. Ballardini, V. Balzani, A. Credi, M.T. Gandolfi, S. Menzer, L. Pérez-Garcia, L. Prodi, J.F. Stoddart, M. Venturi, A.J.P. White and D.J. Williams, *J. Am. Chem. Soc.*, 1995, **117**, 11171.

able to be isomerised photochemically within a free tetracationic cyclophane, unfortunately, attempts to induce such a configurational change within the corresponding [2]-catenanes failed. It appears that the inability to photoaddress the latter is associated with the inhibiting effect of low energy charge-transfer bands in these assemblies.

Finally, it is noted that photochemical and electrochemical investigations on other aromatic-containing catenane systems have also been reported.[31,32]

5.3.1 Bis-[2]-catenanes

In order to extend their studies to more complex systems, the Stoddart group directed their efforts towards the synthesis of the bis-[2]-catenane **16**.[33] The approach

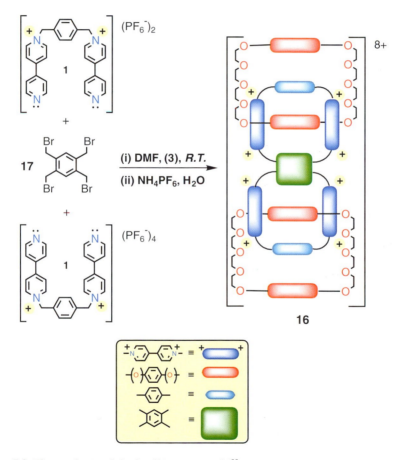

Figure 5.9 *The synthesis of the bis-[2]-catenane* **16**[33]

[31] R. Ballardini, V. Balzani, A. Credi, M.T. Gandolfi, L. Prodi, M. Venturi, L. Pérez-Garcia and J.F. Stoddart, *Gazz. Chim. Ital.*, 1995, **125**, 353.

[32] M. Bauer, W.M. Müller, U. Müller, K. Rissanen and F. Vögtle, *Liebigs Ann.*, 1995, 649.

[33] P.R. Ashton, A.S. Reder, N. Spencer and J.F. Stoddart, *J. Am. Chem. Soc.*, 1993, **115**, 5286.

employed was based on the use of the four reactive benzylic groups of 1,2,4,5-tetrakis(bromomethyl)benzene to construct two fused cyclo-bis(paraquat-*p*-phenylene) rings – each catenated with a bis-*p*-phenylene-34-crown-10 macrocycle (Figure 5.9). The self-assembly process in this case involved the reaction of two molar equivalents of the dication **1** with one molar equivalent of the tetrabromo compound **17** in dry *N,N*-dimethylformamide or acetonitrile over 10 days at room temperature in the presence of an excess of the dibenzo-crown macrocycle **3**. The chiral product was obtained in 13% yield as a deep red crystalline solid.

The above successful incorporation of the tetra-substituted benzene ring to produce the fused tetracationic cyclophane units, relies on being able to react the former in a para/para fashion rather than in an ortho/ortho or meta/meta one. The present product represents the successful interlinking of three major components. If one considers that a one-step synthetic procedure was employed, then it is a total of five molecular species that are, in fact, involved in the self-assembly process.

With this system two translational isomers are possible (Figure 5.10) and, from the proton NMR spectrum, both are seen to be populated to an equal extent at −70 °C in deutero-acetone. Under these conditions, the movement of the

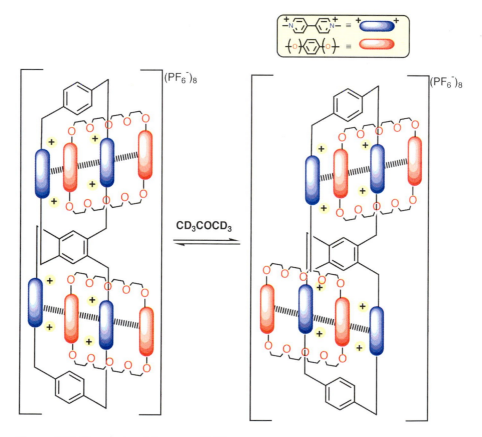

Figure 5.10 *Translational isomers of* **16**[33]

bis-*p*-phenylene-34-crown-10 rings with respect to the pairs of bipyridinium units is slow on the ¹H NMR timescale, while a time-averaged spectrum is obtained at higher temperatures. The difunctional nature of this system may be seen as a step towards the additional complexity that will most certainly be a feature of future supramolecular devices.

The self-assembly of other bis-[2]-catenanes have been reported.[34,35] For example, a second bis-[2]-catenane **18** showing a somewhat different architecture (in which a flexible link is incorporated between each catenane moiety) has been prepared in overall 31% yield (Figure 5.11).[34] Assembly of this species again takes place under mild conditions (14 days at room temperature in acetonitrile). The self-assembly of other bis-[2]-catenanes has also been reported.

Other 'expanded' catenanes, **19** and **20**, each incorporating a bis-macrocyclic

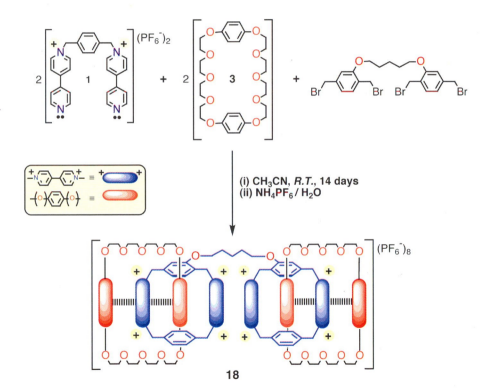

Figure 5.11 *Synthesis by self-assembly of the bis-[2]-catenane* **18** *in which the components are linked by a flexible chain*[34]

34 P.R. Ashton, J.A. Preece, J.F. Stoddart and M.S. Tolley, *Synlett*, 1994, 789.
35 P.R. Ashton, J. Huff, S. Menzer, I.W. Parsons, J.A. Preece, J.F. Stoddart, M.S. Tolley, A.J.P. White and D.J. Williams, *Chem. Eur. J.*, 1996, **2**, 31.

component whose synthesis was based on the tetravalency of the tetrathiafulvalene-2,3,6,7-tetrathiolate unit, have been reported.[36]

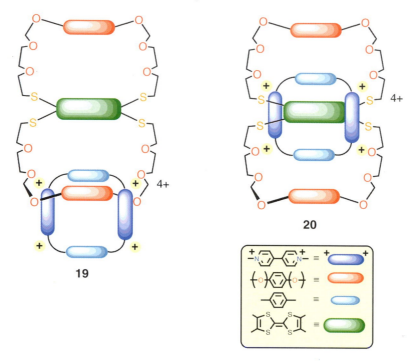

On incorporating two dioxyphenylene groups in the bis-macrocycle, both the *cis* and *trans* isomers (with respect to the substitution on the central tetrathiafulvalene unit) of the corresponding expanded catenane were obtained. However, replacement of the dioxyphenylenes with 9,10-dioxyanthrylenes resulted in the selective self-assembly of only the *trans* isomer of the corresponding catenane. In a subsequent study, a 'clipping' reaction (under ultra high pressure) was employed to assemble the interesting catenane **21** – this time incorporating three tetrathiafulva-

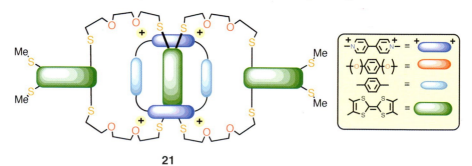

36   Z.-T. Li, P.C. Stein, N. Svenstrup, K.H. Lund and J. Becher, *Angew. Chem., Int. Ed. Engl.*, 1995, **34**, 2524; J. Becher, Z.-T. Li, P. Blanchard, N. Svenstrup, J. Lau, M.B. Nielsen and P. Leriche, *Pure Appl. Chem.*, 1997, **69**, 465.

lene units.[37] In contrast to the above, the *cis* isomer was isolated exclusively (in 20% yield) in this case. The adoption of a *cis* arrangement by the central tetrathiafulvalene moiety perhaps reflects the formation by the peripheral tetrathiafulvalenes of π–π donor–acceptor interactions with the two-bipyridinium rings, thus resulting in configurational control during the assembly process.

5.3.2 Higher-order Catenanes

The macrocyclic polyether, tris(1,5-naphtho)-57-crown-15 (Figure 5.12) contains a sufficiently large ring size to permit the simultaneous accommodation of two threaded 4,4′-bipyridinium units, which are then appropriately oriented for further template formation of tetracationic cyclophanes.[38] However, the efficacy of template formation for [3]-catenane formation proved to be less than anticipated.[11d,39] The

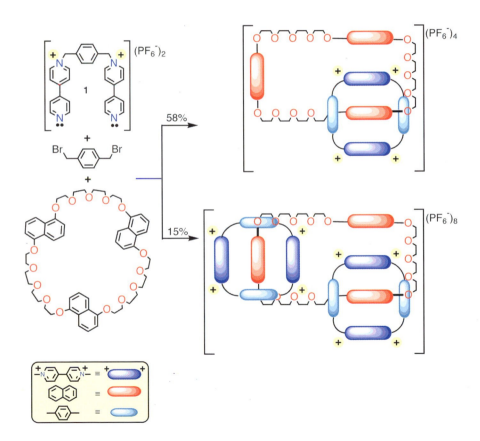

Figure 5.12 *Formation of a [3]-catenane from tris(1,5-naphtho)-57-crown* **15**[38]

37 Z.-T. Li and J. Becher, *Chem. Commun.*, 1996, 639.
38 D.B. Amabilino, P.R. Ashton, J.F. Stoddart, S. Menzer and D.J. Williams, *J. Chem. Soc., Chem. Commun.*, 1994, 2475.

respective yields of the corresponding [2]- and [3]-catenane products, obtained in a one-pot system, were 58 and 15%, respectively. Hence, there is a reluctance for the precursor [2]-catenane to react further. A clue to this reluctance was obtained from the solid state structure of this catenane, which showed that all of the aromatic subunits were involved in intramolecular π–π stacking interactions (with intermolecular π-stacking also in evidence). It appears likely that formation of the [3]-catenane in solution is inhibited by the above arrangement and associated steric factors and/or charge–charge repulsions involving the above π-electron-rich units of the macrocyclic polyether.

In contrast, when a larger tetracationic cyclophane ring was employed in another set of experiments, the formation of new [3]-catenanes was found to be more facile.[40] The expanded ring was obtained by replacing the phenylene linker units in the 'normal' tetracationic cyclophane with biphenylene units. This larger ring was found to form readily around two threaded crowns when the latter are present in the acetonitrile reaction mixture in several-fold excess. Both bis-*p*-phenylene-34-crown-10 and 1,5-dinaphtho-38-crown-10 were successfully employed in separate reactions to produce the corresponding [3]-catenanes in 20 and 31% yields, respectively.

There has now been a range of studies aimed at extending the self-assembly process to the synthesis of other [*n*]-catenanes.[41]

The consequences of increasing the size of the polyether macrocycle to 68 members (that is, by a factor of two over the small ring system) was investigated. Accordingly, the synthesis of tetrakis-*p*-phenylene-68-crown-20 was achieved in dimethylformamide at room temperature in excellent yield and this, in turn, was initially used to produce the [2]-catenane **22** (Figure 5.13).[42] From the temperature dependence of the ^{1}H NMR spectrum, **22** appears to behave like a *molecular train* – with the tetracationic cyclophane travelling from 'station' to 'station' around the

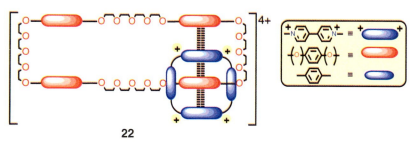

22

Figure 5.13 *A molecular train set*[42]

[39] It is noted that the synthesis of [3]-catenanes in low yield *via* multistep, direct procedures had also been achieved quite early: G. Schill and C. Zürcher, *Angew Chem., Int. Ed. Engl.,* 1969, **8**, 988; G. Schill, K. Murjahn and W. Vetter, *Justus Liebigs Ann. Chem.,* 1970, **740**, 18; G. Schill, K. Rißler, H. Fritz and W. Vetter, *Angew. Chem., Int. Ed. Eng.,* 1981, **20**, 187; G. Schill, W. Beckmann and H. Fritz, *Chem. Ber.,* 1982, **115**, 2683.

[40] P.R. Ashton, C.L. Brown, E.J.T. Chrystal, T.T. Goodnow, A.E. Kaifer, K.P. Parry, A.M.Z. Slawin, N. Spencer, J.F. Stoddart and D.J. Williams, *Angew Chem., Int. Ed. Engl.,* 1991, **30**, 1039.

[41] D.B. Amabilino, P.R. Ashton, C.L. Brown, E. Córdova, L.A. Godinez, T.T. Goodnow, A.E. Kaifer, S.P. Newton, M. Pietraszkiewicz, D. Philp, F.M. Raymo, A.S. Reder, M.T. Rutland, A.M.Z. Slawin, N. Spencer, J.F. Stoddart and D.J. Williams, *J. Am. Chem. Soc.,* 1995, **117**, 1271.

[42] P.R. Ashton, C.L. Brown, E.J.T. Chrystal, K.P. Parry, M. Pietraszkiewicz, N. Spencer and J.F. Stoddart, *Angew. Chem., Int. Ed. Engl.,* 1991, **30**, 1042.

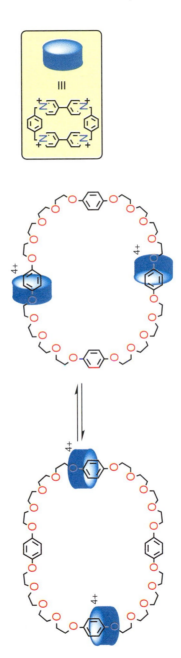

Figure 5.14 A [3]-catenane in which the two cyclophane units cycle between four 'stations' while remaining diametrically opposed[42]

loop. At room temperature the 'train' travels from one 'station' to the next about 300 times a second. Further, when the reaction conditions were modified and the reaction carried out at 10 kbar, then the corresponding [3]-catenane was formed. The dynamic ^{1}H NMR spectra in this case indicate that the two 'trains' travel in unison around the polyether 'circle line' (Figure 5.14), keeping a one 'station' separation between them. The time spent travelling between stations is similar to that found for the single train case.[1]

Based on the chemistry discussed so far, two-step procedures for the assembly of [4]- and [5]-catenanes have also been developed.[43] In part, the success of these syntheses reflects the use of the 'intermediate ring size' crown, tris-*p*-phenylene-[51]-crown-15 which provides two non-equivalent templating sites for the catenation reactions. The strategies employed for the construction of these extraordinary multi-linked systems are given in Figure 5.15. The respective products were characterised by means of FAB-MS and LSI-MS, with the fragmentation patterns

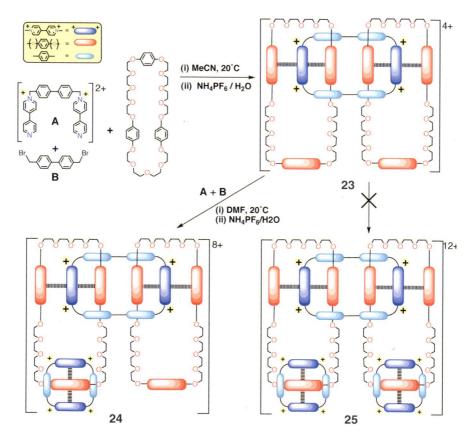

Figure 5.15 *The two-step synthesis of the [4]-catenane **24** in 31% yield via the [3]-catenane **23**. In contrast, the yield of the corresponding [5]-catenane **25** was minuscule*[43]

[43]   D.B. Amabilino, P.R. Ashton, A.S. Reder, N. Spencer and J.F. Stoddart, *Angew. Chem., Int. Ed. Engl.*, 1994, **33**, 433.

reflecting the breaking of successive links in each catenane 'chain'. While the yield of [4]-catenane **24** was 22%, that of the [5]-catenane **25** was negligible – at less than 0.5%.

Although the precise reason for the reluctance of the [4]-catenane **24** to add a fifth ring to yield **25** was unclear, it was reasoned that it might be associated with the recognition and binding properties of the macrocyclic ether component. Since derivatives of 1,5-dihydroxynaphthalene are more strongly bound within a cyclo-bis(paraquat-*p*-phenylene) ring[27] than in the analogous hydroquinone derivative, it was decided to replace the hydroquinone residues in the tris-*p*-phenylene[51]crown-15 ring of **23** with 1,5-dioxynaphthalene units.[44] This strategy proved successful – with the two-step self-assembly process proceeding at ambient temperature and pressure to yield the required linear array of five interlocked rings. The latter was named *olympiadane* after the Olympic symbol (Figure 5.16). In this case the violet product was able to be isolated in 5% yield even though, because of incomplete reaction, more of the corresponding [4]-catenane (31% yield) was obtained.

Once again, the successful synthesis of this remarkable molecule reflects the ability to exploit and manipulate the weak π–π and hydrogen bonding interactions between ring components as they assemble – a process learnt from the numerous experiments on simpler systems described so far. Further, preliminary evidence for the formation of corresponding [6]- and [7]-catenanes was also obtained in this study; this thus provided a pointer towards further synthetic targets – which were subsequently achieved.[45] Namely, when the tricatenane **23** was treated with **26** and **27** (Figure 5.17) in *N,N*-dimethylformamide at 12 kbar for 6 days, then the even more remarkable [7]-catenane **28** was isolated in 26% yield (after chromatography) along with the [5]-catenane **29** (30% yield) and the [6]-catenane **30** in 28% yield.[46]

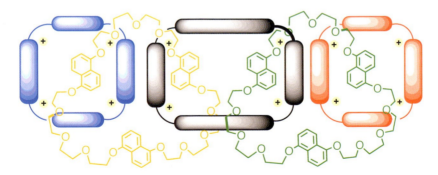

Figure 5.16 *The multiply-linked catenane, olympiadane*[44]

44   D.B. Amabilino, P.R. Ashton, A.S. Reder, N. Spencer and J.F. Stoddart, *Angew. Chem., Int. Ed. Engl.*, 1994, **33**, 1286.
45   D.B. Amabilino, P.R. Ashton, S.E. Boyd, J.Y. Lee, S. Menzer, J.F. Stoddart and D.J. Williams, *Angew. Chem., Int. Ed. Engl.*, 1997, **36**, 2070.
46   D.B. Amabilino, P.R. Ashton, V. Balzani, S.E. Boyd, A. Credi, J.Y. Lee, S. Menzer, J.F. Stoddart, M. Venturi and D.J. Williams, *J. Am. Chem. Soc.*, 1998, **120**, 4295.

The X-ray structure of the deep violet [7]-catenane confirmed its structure and indicated that all recognition sites are beautifully involved in mutually compatible π-stacking interactions.

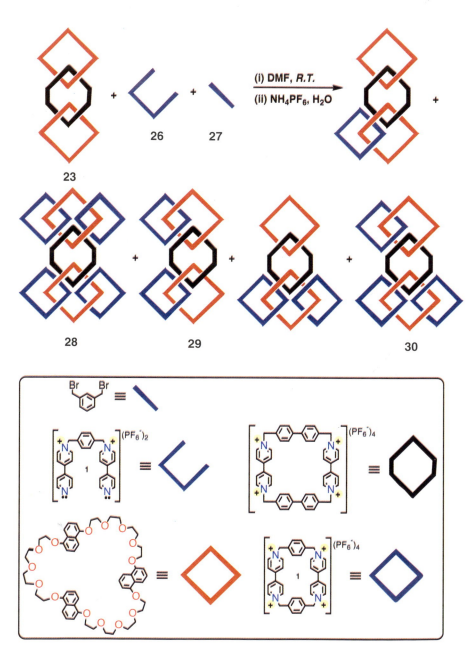

Figure 5.17 *The multistep template-directed synthesis of the high-order catenanes* **28–30**[43]

5.4   Porphyrin-containing Catenanes

In other work, the principles discussed above have been applied to the template production of porphyrin-containing [2]-catenanes with, once again, the assembly process being based primarily on π–π interactions but including contributions from both C–H⋯O and C–H⋯π hydrogen bonding.[47,48]

The incorporation of a porphyrin group in a catenane raises the prospect that the former might serve as a centre at which photochemical, electrochemical or chemical processes are used to control the dynamic and/or conformational behaviour of the assembly as a whole. In addition, the tuning of such processes should be possible through the choice of suitable metallation as well as through the ability to vary substituents on the periphery of the porphyrin ring.

Initially, a number of hydroquinone-containing ether chains were synthesised which were subsequently 'strapped' across a porphyrin ring (see Figure 5.18). Individual strapped porphyrins of this type were demonstrated to complex fairly strongly with paraquat (K values of 1955 and 1640 dm^3 mol^{-1} were determined for the complexes of **31** and **32**, respectively) with, as expected, parallel orientations of the hydroquinone derivative, paraquat and the porphyrin ring occurring in these complexes. Based on these experiments, the template-directed, self-assembly of the first examples of porphyrin-containing [2]-catenanes were synthesised (Figure 5.19). The zinc-containing products **33** and **34** were isolated in 20 and 28% yield, respectively, with the zinc-free derivatives **35** and **36** being generated from them by treatment with aqueous hydrochloric acid followed by neutralisation and anion exchange. All products were characterised by FAB-MS and investigated using ^{1}H NMR. In all cases the overall orientation of the tetracationic cyclophane, with bipyridinium and porphyrin rings parallel, is retained in solution. For these hydroquinone-containing [2]-catenanes, the bipyridinium subunits in the respective tetracations were observed to undergo exchange between their 'inside' and 'outside' environments (corresponding to rotation about the hydroquinone ring axis), with the rate of rotation strongly influenced by the length of the ether strap present.

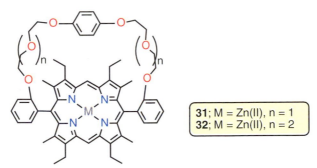

31; M = Zn(II), n = 1
32; M = Zn(II), n = 2

Figure 5.18 *Strapped porphyrins* **31** *and* **32**[48]

[47]   M.J. Gunter and M.R. Johnston, *J. Chem. Soc., Chem. Commun.*, 1992, 1163.
[48]   M.J. Gunter, D.C.R. Hockless, M.R. Johnston, B.W. Skelton and A.H. White, *J. Am. Chem. Soc.*, 1994, **116**, 4810.

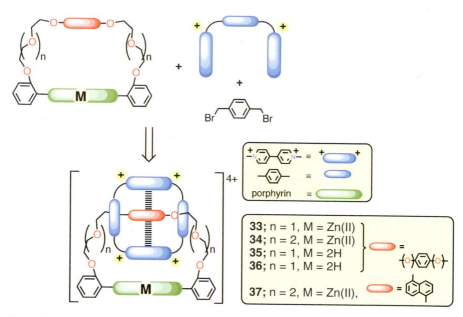

Figure 5.19 *The template-directed, self-assembly of porphyrin-containing [2]-catenanes*[48]

The general procedure was also applied to the syntheses of the related naphthalene-containing [2]-catenane **37**, which was obtained in 45% yield. (It is worth noting that, in a separate experiment, a very high association constant of 21 100 dm^3 mol^{-1} at 25 °C was observed for the binding between the polyether macrocycle and paraquat.) The successful incorporation of the strong π-electron-donating, 1,5-substituted naphthoquinol group in the new catenane – starting from the corresponding macrocycle containing triethylene glycol links between the porphyrin and nathphalene π-electron centres – provides evidence for the generality of the synthesis of porphyrin catenanes of this type. However, it needs to be noted that the smaller ring macrocycle, incorporating diethylene glycol chains in the porphyrin strap, is now too small to allow such catenane formation.

Tetracation exchange processes (ring rotation) similar to those for the above hydroquinone catenanes were documented for the naphthalene system together with an additional process corresponding to the decomplexation/complexation of the naphthalene ring (designated as 'out, turn around, and in again'). The latter is related to that observed by Stoddart and coworkers for the 'non-porphyrin', naphthalene-containing [2]-catenanes.[20] In essence, it is assigned to a process that involves the loss of the naphthalene ring from within the tetracation, reorientation of the ring, followed by recomplexation. Comparison of the 'ring rotation' processes in the hydroquinone and naphthalene systems indicates that the process is slower in the latter case. This is in accordance with the presence of a stronger π-interaction between the naphthalene ring and the tetracation.

Demetallated porphyrin catenanes of the type just discussed have been demon-

strated to undergo protonation in strongly acid solution.[49] For each system, protonation results in electrostatic repulsion between the two-catenane components and,
as a consequence, both conformational reorientations and changes in rotation rates
are observed. Acid–base chemistry thus provides yet another means of inducing
predictable movement of the mechanically-linked components – adding to the
armoury of potential mechanisms for controlling molecular scale devices.

5.5 Neutral Catenanes

In early studies two different groups were successful in producing related uncharged
[2]-catenanes incorporating amide linkages. In each case the systems consisted of
interlocked macrocycles of similar structure held together in well-defined arrangements by a combination of hydrogen bonding and π-interactions; once again these
same interactions act as templates during formation of the interlocked rings. The
respective systems are composed of interlocked macrocycles of type (**38a**, R = H)[50]
and (**38b**, R = OCH₃).[51] More recently, related systems based on **39**[52] and **40**[53] have
been reported.

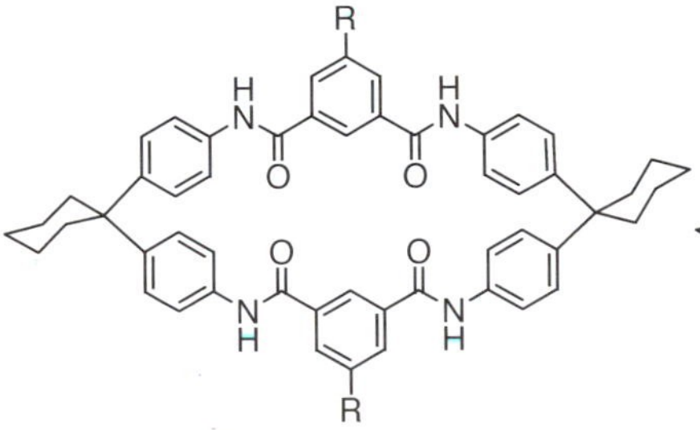

38a; R = H
38b; R = OCH₃

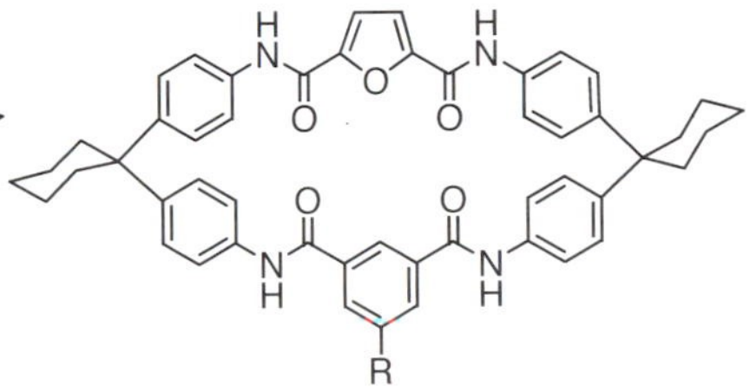

39

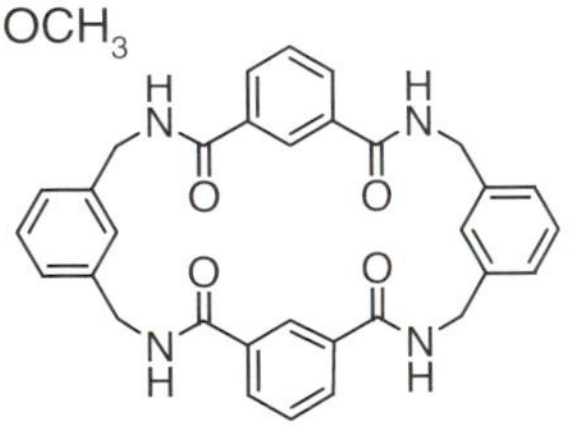

40

The first of the above [2]-catenane systems was prepared by a one-pot double-
macrocyclisation reaction to give the linked product in 34% yield, with the other
(non-catenated) products from this reaction being the corresponding single and

49   M.J. Gunter and M.R. Johnston, *J. Chem. Soc., Chem. Commun.*, 1994, 829.
50   C.A. Hunter, *J. Am. Chem. Soc.*, 1992, **114**, 5303.
51   F. Vögtle, S. Meier and R. Hoss, *Angew. Chem., Int. Ed. Engl.*, 1992, **31**, 1619.
52   S. Ottens-Hildebrandt, M. Nieger, K. Rissanen, J. Rouvinen, S. Meier, G. Harder and F. Vögtle, *J. Chem. Soc., Chem. Commun.*, 1995, 777.
53   A.G. Johnston, D.A. Leigh, R.J. Pritchard and M.D. Deegan, *Angew. Chem., Int. Ed. Engl.*, 1995, **34**, 1209.

double [2+2] rings. The dynamic behaviour of the [2]-catenane has been investigated using variable temperature NMR. Interestingly, the bulky cyclohexyl groups prevent free rotation of the macrocycles and hence, exchange of the inside and outside parts of the catenane is inhibited. Even so, the molecule is not completely rigid and has been demonstrated to switch between two enantiomeric conformations at a room temperature rate of 1 s⁻¹. The solid state X-ray structure of the above [2]-catenane has been determined (Figure 5.20).[54] In each macrocycle, one isophthaloyl subunit has its amide groups aligned *cis* to one another, forming a pocket for hydrogen bond formation to the carbonyl from the other macrocycle.

As expected, the related [2]-catenane derived from the substituted derivative (**38b**, R = OCH₃), prepared independently by the Vögtle group,[51] in general shows related properties to its unsubstituted parent investigated by Hunter.[50] In these systems the perpendicular preorganisation of the catenane building blocks appears to reflect three templating influences: steric complementarity, hydrogen bonding between carbonyl oxygen atoms and amide protons, and possibly π-interactions between the aryl rings of the host and guest subunits.

The studies have been extended to the production of 'intermediate' systems containing different combinations of interlinked singly-substituted and unsubstituted rings of the above type.[55] For example, using the alternative synthetic procedures shown in Figure 5.21, it was possible to isolate three of the four possible isomeric catenanes illustrated. The respective mixtures were separated chromatographically, confirming that there is hindered circumrotation of the catenane rings.

The [2]-catenane derived from the furano-containing macrocycle **41** was demonstrated to exist as a mixture of isomers of the types **42** and **43**.[52,55] No evidence for

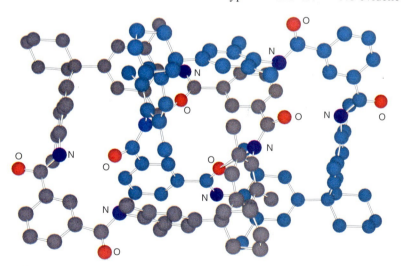

Figure 5.20 *[2]-Catenane prepared by a one-pot double-macrocyclisation reaction;[54] each macrocycle is related by crystallographic symmetry*

54   H. Adams, F.J. Carver and C.A. Hunter, *J. Chem. Soc., Chem. Commun.*, 1995, 809.
55   F. Vögtle, R. Jager, M. Handel and S. Ottens-Hildebrandt, *Pure Appl. Chem.*, 1996, **68**, 225.

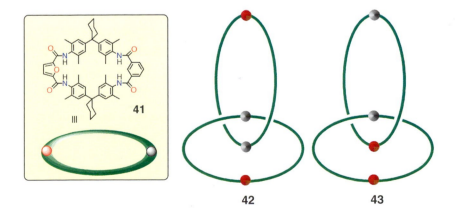

Figure 5.21 *The route-dependent preparation of isomeric catenanes*[55]

the formation of the remaining isomer – in which the furano groups are adjacent – was obtained. The structure of the first of these isomers was confirmed by X-ray structure analysis, enabling details of the interaction between the macrocyclic rings to be elucidated. Both interlocked rings have identical conformations, with the rings associated with numerous intra- and inter-molecular hydrogen bonds. Each isophthaloyl group is 'buried' in the cavity of the opposite macrocycle and appears to be involved in inter-ring π-stacking. One of the carbonyl groups is hydrogen bonded to the amide proton of the second isophthaloyl group. The other carbonyl takes part in a network of hydrogen bonds connecting the two-amide protons of adjacent furanoyl units as well as the furan oxygen.

Related to the above, the one step, eight-molecule condensation shown in Figure 5.22 yields the corresponding [2]-catenane **44**.[53] The product consists of interlinked 26-membered rings making this the smallest interlocked ring system of this type then prepared. The X-ray structure of the product is once again in accordance with catenane formation being largely controlled by hydrogen bonding between the 1,3-diamide units and the carbonyl groups on the acid chloride (or other intermediates) – with possible assistance from π-stacking interactions between the electron-rich xylylene moieties and the electron-poor isophthaloyl rings.

A number of related catenanes,[56] including systems assembled directly from commercially available precursors[57] and a system based on sulfonamide linkages,[58] have also been described.

A structure of the latter type has been employed as a precursor for the construction of the first pretzel-shaped assembly *via* the conventional bridging dialkylation reaction shown in Figure 5.23.[59] The success of this strategy rests upon the

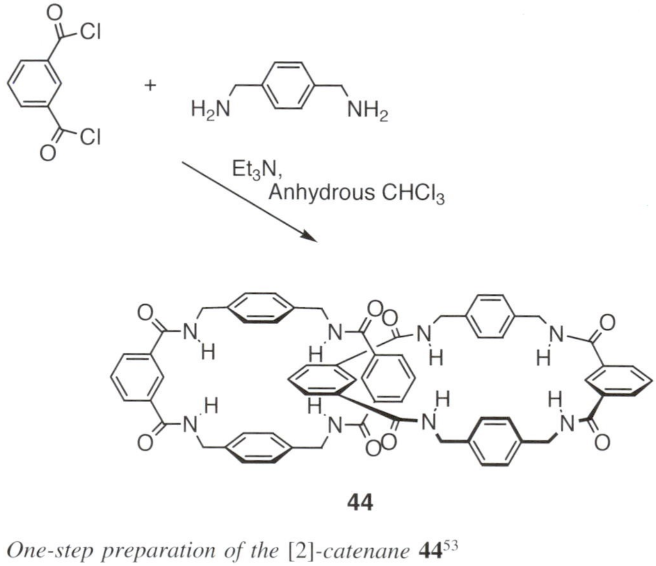

44

Figure 5.22 *One-step preparation of the [2]-catenane* **44**[53]

56   F.J. Carver, C.A. Hunter and R.J. Shannon, *J. Chem. Soc., Chem. Commun.*, 1994, 1277; D.A. Leigh, K. Moody, J.P. Smart, K.J. Watson and A.M.Z. Slawin, *Angew. Chem., Int. Ed. Engl.* 1996, **35**, 306; S. Ottens-Hildebrandt, S. Meier, W. Schmidt and F. Vögtle, *Angew. Chem., Int. Ed. Engl.*, 1994, **33**, 1767.
57   A.G. Johnston, D.A. Leigh, L. Nezhat, J.P. Smart and M.D. Deegan, *Angew. Chem., Int. Ed. Engl.*, 1995, **34**, 1212.
58   S. Ottens-Hildebrandt, T. Schmidt, J. Harren and F. Vögtle, *Liebigs Ann.*, 1995, 1855.
59   R. Jager, T. Schmidt, D. Karbach and F. Vögtle, *Synlett*, 1996, 723; C. Yamamoto, Y. Okamoto, T. Schmidt, R. Jager and F. Vögtle, *J. Am. Chem. Soc.*, 1997, **119**, 10547 and references therein.

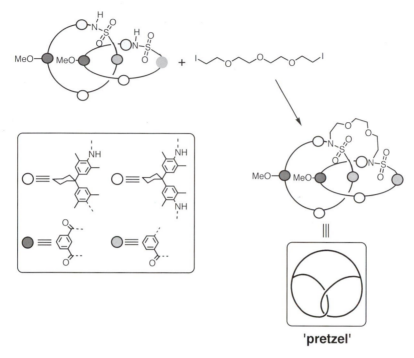

Figure 5.23 *The synthesis of a pretzel-shaped molecule* via *the covalent linkage of the components of a [2]-catenane*[59]

relative acidities of sulfonamide *versus* amide groups: owing to their greater acidity, sulfonamide groups can be selectively deprotonated even in the presence of amide groups. This enables the selective alkyl substitution of the respective rings, as illustrated in the figure, to be carried out.

5.6 Cyclodextrin-containing Systems

In an attempt to increase the number of useful 'components' for construction of new catenanes, the Stoddart group recognised that cyclic sugars of the cyclodextrin type were promising candidates.[60]

As mentioned in Chapter 4, cyclodextrins are cyclic glucopyranose oligomers having at least 6 (and up to 13) monomeric units linked by α-(1,4) linkages. As also mentioned, these water-soluble, cone-shaped molecules have been well documented to include an extremely large number of both inorganic and organic guests in their hydrophobic cavities.

Since they anticipated difficulties in the isolation of unmodified catenated cyclodextrins, the Stoddart group[60] initiated their studies using the poly-methylated derivative **45** of the seven-membered sugar unit, β-cyclodextrin. The latter has

[60]   D. Armspach, P.R. Ashton, C.P. Moore, N. Spencer, J.F. Stoddart, T.J. Wear and D.J. Williams, *Angew. Chem., Int. Ed. Engl.*, 1993, **32**, 854.

good solubility in both organic solvents and water. Further, the torus-shaped cavity in this derivatised cyclodextrin is somewhat more 'rigid' than occurs in the parent ring; this extra rigidity is expected to enhance the binding of the former to aromatic ring systems. The reaction of the water-soluble inclusion complex formed between equimolar amounts of cyclodextrin **45** and the long-chain polyether diamine **46** (in the presence of base) with 1.1 equivalents of terephthaloyl chloride – as the ring-closing, diacylating reagent – yielded the products **47–52** (Figure 5.24). The single ring macrocycles **47** and **48** were obtained in 12 and 3.5% yield, respectively, while the yields were 3 and 0.8% for the related [2]-catenanes, **49** and **50**. The isomeric [3]-catenanes **51** and **52** were also obtained as an equimolar mixture in overall 1.1% yield. Subsequent studies using similar procedures (again under aqueous conditions) have led to a wide range of other derivatives, but in all cases the yields were once again very low.[61] Clearly, for these systems any templating effects present do not efficiently promote all steps of the reaction sequence.

5.7 DNA-based Systems

In highly novel studies, Seeman and co-workers[62] have been successful in constructing elaborate structures from DNA strands. For example, the solution phase preparation of a covalently closed cube-like molecular complex, incorporating twelve double-helical edges of equal length arranged around eight vertices, has been achieved. Each of the six faces is defined by a single stranded cyclic molecule, doubly catenated to four neighbouring strands. The product corresponds to a [6]-catenane and represents the first example of a closed polyhedral object being constructed from DNA.

In a further study, a solid support has been employed to assemble a closed [14]-catenane (again from DNA) whose double helical edges show the connectivity of a truncated octahedron.[63] The structure is composed of 14 cyclic DNA molecules and contains 2550 nucleotides; it has a molecular weight of about 790 000 Daltons. The successful assembly of structures such as these owes much to the use of biochemical techniques and, in particular, the application of restriction enzymes to cleave DNA strands at appropriate places and a DNA ligase to join the 'sticky' ends where required. Overall, studies of this latter type represent a remarkable achievement and suggest that DNA is a tractable medium for nanoscale construction. However, the challenge remains to make like structures from more robust materials than DNA.

5.8 An Immobilised System

Catenanes have been constructed on a gold surface using the self-assembly sequence shown in Figure 5.25.[64] The confinement of the cyclic tetra-cations at the gold sur-

[61] D. Armspach, P.R. Ashton, R. Ballardini, V. Balzani, A. Godi, C.P. Moore, L. Prodi, N. Spencer, J.F. Stoddart, M.S. Tolley, T.J. Wear and D.J. Williams, *Chem. Eur. J.* 1995, **1**, 33.
[62] J. Chen and N.C. Seeman, *Nature*, 1991, **350**, 631; N.C. Seeman, J.H. Chen, S.M. Du, J.E. Mueller, Y.W. Zhang, T.J. Fu, Y.L. Wang, H. Wang and S.W. Zhang, *New. J. Chem.*, 1993, **17**, 739.
[63] Y. Zhang and N.C. Seeman, *J. Am. Chem. Soc.*, 1994, **116**, 1661.

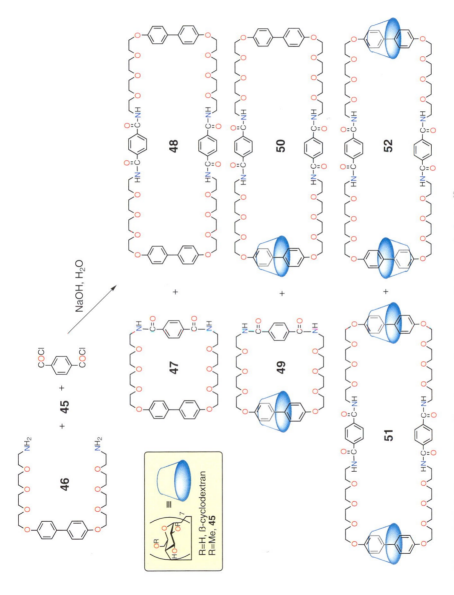

Figure 5.24 *Self-assembled [2]-catenanes based on methylate cyclodextrin*[60]

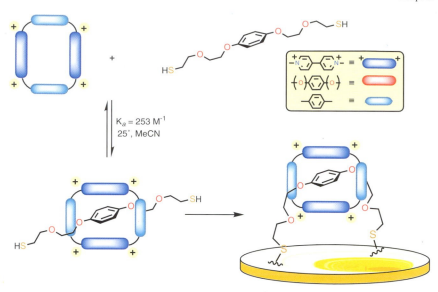

Figure 5.25 *Synthesis of an immobilised catenane on a gold substrate*[64]

face was confirmed using electrochemical methods; nevertheless, these also indicated that the catenane groups covered only a small fraction of the surface. However, the result provides a clear demonstration that a two-component supramolecular structure is able to be assembled on a metal surface – a result of potential significance for new sensor development.

5.9 Other Catenanes and Knots

Many other catenane species have been synthesised. These range from further simple [2]-catenanes,[65] other [2]- as well as [3]-catenane species incorporating redox-active tetrathiafulvalene groups,[66] other multi-ring derivatives,[67,68] through to polymeric species incorporating catenane domains as part of a larger macromole-

[64] T. Lu, L. Zhang, G.W. Gokel and A.E. Kaifer, *J. Am. Chem. Soc.*, 1993, **115**, 2542.
[65] See for example: W. Littke and E. Logemann, *J. Math. Chem.*, 1993, **13**, 53; M. Nakagawa, M. Rikukawa, K. Sanui and N. Ogata, *Synth. Mets.*, 1997, **86**, 1873; R. Ballardini, V. Balzani, A. Credi, M.T. Gandolfi, D. Marquis, L. Perez-Garcia and J.F. Stoddart, *Eur. J. Org. Chem.*, 1998, 81; P.R. Ashton, S.E. Boyd, S. Menzer, D. Pasini, F.M. Raymo, N. Spencer, J.F. Stoddart, A.J.P. White, D.J. Williams and P.G. Wyatt, *Chem. Eur. J.*, 1998, **4**, 299; Z.T. Li, G.Z. Ji, S.D. Yuan, A.L. Du, H. Ding and M. Wei, *Tetrahedron Lett.*, 1998, **39**, 6517; S. Capobianchi, G. Doddi, G. Ercolani and P. Mencarelli, *J. Org. Chem.*, 1998, **63**, 8088; D.G. Hamilton, N. Feeder, L. Prodi, S.J. Teat, W. Clegg and J.K.M. Sanders, *J. Am. Chem. Soc.*, 1998, **120**, 1096.
[66] Z.-T. Li, P.C. Stein, J. Becher, D. Jensen, P. Mork and N. Svenstrup, *Chem. Eur. J.*, 1996, **2**, 624; R. Ballardini, V. Balzani, A. Credi, C.L. Brown, R.E. Gillard, M. Montalti, D. Philp, J.F. Stoddart, M. Venturi, A.J.P. White, B.J. Williams and D.J. Williams, *J. Am. Chem. Soc.*, 1997, **119**, 12503; M. Asakawa, P.R. Ashton, V. Balzani, A. Credi, C. Hamers, G. Mattersteig, M. Montalti, A.N. Shipway, N. Spencer, J.F. Stoddart, M.S. Tolley, M. Venturi, A.J.P. White and D.J. Williams, *Angew. Chem., Int. Ed. Engl.*, 1998, **37**, 333.

cular architecture.[68,69] Exotica, such as a teracationic catenane derivative incorporating a fullerene covalently bound to an interlinked (uncharged) crown ring component[70] have also been reported.

The Stoddart group has used the principles employed in their catenane studies to synthesise both a molecular trifoil knot and its isomeric large ring in very low yield.[71] These isomers were separated by HPLC and characterised by liquid secondary ion mass spectrometry. The synthetic procedure, which is illustrated schematically in Figure 5.26, was based on the formation of a double-stranded supramolecular complex between an acyclic π-electron-rich, 1,5-dioxynaphthalene-based polyether and an acyclic π-electron-deficient bipyridinium-based tetracation. The extremely low efficiency of the synthesis in this case appears to reflect an unfavourable solution geometry of the intermediate host–guest complex – the latter

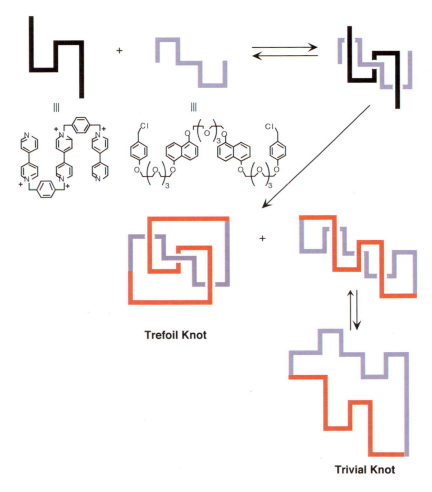

Trefoil Knot

Trivial Knot

Figure 5.26 *Schematic representation of the strategy for obtaining a trefoil knot and its isomeric trivial knot[70]*

favouring inter- rather than intra-complex covalent bond formation in the final steps. Thus polymer formation is promoted over the desired ring-closing reactions.

As mentioned already, a further group of catenanes have been synthesised *via* metal-ion template procedures; discussion of these species is presented in the next chapter as part of a wider treatment of the use of metals in self-assembly processes.

[67]   See for example: D.B. Amabilino and J.F. Stoddart, *Chem. Rev.*, 1995, **95**, 2725; F.M. Raymo and J.F. Stoddart, *Pure Appl. Chem.*, 1996, **68**, 313; M. Asakawa, P.R. Ashton, C.L. Brown, M.C.T. Fyfe, S. Menzer, D. Pasini, C. Scheuer, N. Spencer, J.F. Stoddart, A.J.P. White and D.J. Williams, *Chem. Eur. J.*, 1997, **3**, 1136; D.B. Amabilino, P.R. Ashton, J.F. Stoddart, A.J.P. White and D.J. Williams, *Chem. Eur. J.*, 1998, **4**, 460.

[68]   S. Menzer, A.J.P. White, D.J. Williams, M. Belohradský, C. Hamers, F.M. Raymo, A.N. Shipway and J.F. Stoddart, *Macromolecules*, 1998, **31**, 295.

[69]   See for example: H.W. Gibson, M.C. Bheda and P.T. Engen, *Prog. Polym. Sci.*, 1994, **19**, 843; D. Muscat, A. Witte, W. Kohler, K. Mullen and Y. Geerts, *Macromol. Rapid Commun.*, 1997, **18**, 233 and references therein; C. Hamers, F.M. Raymo and J.F. Stoddart, *Eur. J. Org. Chem.*, 1998, 2109.

[70]   P.R. Ashton, F. Diederich, M. Gomez-Lopez, J.-F. Nierengarten, J.A. Preece, F.M. Raymo and J.F. Stoddart, *Angew. Chem., Int. Ed. Engl.*, 1997, **36**, 1448.

[71]   P.R. Ashton, O.A. Matthews, S. Menzer, F.M. Raymo, N. Spencer, J.F. Stoddart and D.J. Williams, *Liebigs Ann.*, 1997, 2485.

CHAPTER 6

Metal-directed Synthesis – Rotaxanes, Catenanes, Helicates and Knots

6.1 Introduction

There are now many self-assembled supramolecular systems in which metal ions have played an important, if not essential, role in the assembly process. In this chapter, the use of metal-directed synthesis to achieve assembled systems is discussed.

Largely through the early work of Daryle Busch and his group,[1] the possibility of using metal ions for metal–template ligand synthesis – to yield the product as its corresponding metal complex – has long been recognised. Although systems of this type will not be specifically discussed here, it needs to be noted that the lessons learnt in these early studies[2,3] about the role of the metal ion in ligand assembly processes remain pertinent to the present discussion.

In both this chapter and the next, emphasis will be given to the formation of metal-containing supramolecular systems in which the metal aids the assembly process. If one considers the large number of metals that nature has provided, the variety of oxidation states and co-ordination numbers that they may assume, coupled with the often unique spectral, magnetic, redox and/or photochemical properties that many of them exhibit; then the attractiveness of using metals in supramolecular architectures, both as structural elements and as centres of functionality, becomes apparent. It is a lesson that has long been known to Nature. Despite this, only a limited number of metal ions have so far been incorporated into supramolecular systems. The potential for constructing novel metal-containing systems that bridge the gap between classical co-ordination and organic supramolecular chemistry seems virtually unlimited. As such, workers in this area are able to

---

[1] D.H. Busch, *J. Inclusion. Phenom. Mol. Recognit. Chem.*, 1992, **12**, 389; T.J. Hubin, A.G. Kolchinski, A.L. Vance and D.H. Busch, *Adv. Supramolecular Chem.*, 1999, **5**, 237.

[2] L.F. Lindoy, *Quart. Rev.* 1971, XXV, 379.

[3] L.F. Lindoy, *The Chemistry of Macrocyclic Ligand Complexes*, Cambridge University Press, Cambridge, 1989, pp. 1–269.

call upon the vast store of published metal co-ordination and (organometallic) chemistry that has accumulated over the past 150 years or so. It is a rich resource, waiting to be exploited.

6.2   Catenanes and Rotaxanes

6.2.1   Synthesis of a [2]-Catenane

In 1984, Dietrich-Buchecker, Sauvage and Kern[4] described the elegant metal-template synthesis (Figure 6.1) of the first member (**4**) of a new class of

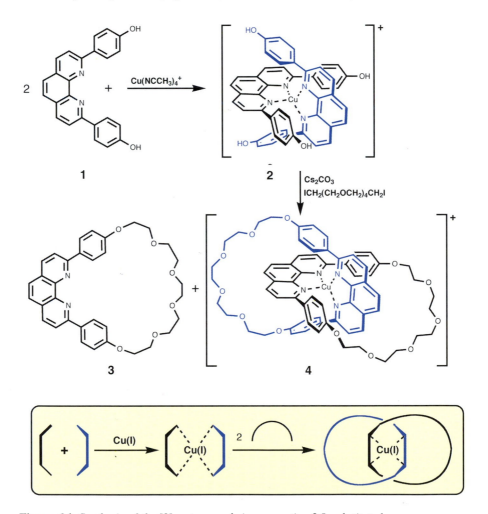

Figure 6.1 *Synthesis of the [2]-catenane,* **4**, *incorporating 2,9-substituted 1,10-phenanthroline ring derivatives co-ordinated to copper(I)[4]*

[4]   C.O. Dietrich-Buchecker, J.-P. Sauvage and J.-M. Kern, *J. Am. Chem. Soc.*, 1984, **106**, 3043.

[2]-catenanes. In this preparation two essentially rigid 2,9-substituted 1,10-phenan-throline derivatives (**1** in Figure 6.1) coordinate in mutually orthogonal positions around a copper(I) ion to yield the corresponding deep red copper(I) complex (**2**). Reaction of this species with 1,4-diiodo-3,6,9,12-tetraoxatetradecane in *N,N*-dimethylformamide under high dilution conditions in the presence of caesium carbonate gave the 'catenate' **4** in 27% yield; the related single-ring macrocycle **3** was also isolated (in its non-complexed form).

In contrast to the behaviour of **2**, attempts to remove the copper ion from the above catenate using electrochemical reduction failed, undoubtedly reflecting the inherent stability of the interlocked complex. However, demetallation of this latter species was achieved chemically by ligand displacement using an excess of the strongly co-ordinating cyanide ion. Subsequent kinetic studies indicated that the cyanide-promoted dissociation of this interlinked ligand complex occurred several orders of magnitude slower than for a non-cyclic ligand analogue; such behaviour almost certainly reflects the necessity to unravel (partially) the two cycles in order for loss of metal to occur in the case of the catenated system.[5]

The assembly of the pseudo-tetrahedral intermediate **2** is clearly the key to the successful formation of the catenane **4**. The process is based on the well documented tendency of copper(I), with its symmetrical d^{10} configuration, to adopt a tetrahedral co-ordination geometry. Indeed, the copper(I) complex of the parent ligand, 1,10-phenanthroline, is a stable 'classical' tetrahedral species that was first investigated in 1933.[6]

Studies of the above type have been extended to the preparation of other single-metal catenanes. These include the synthesis of the nickel(I) and nickel(II) analogues of the above copper(I) catenane **4**; both nickel species were characterised by X-ray diffraction.[7] An electrochemical investigation of the nickel(II)/nickel(I) couple indicated strong stabilisation of the nickel(I) state. This has been attributed to the ready adoption of a favourable tetrahedral geometry by the monovalent nickel cation – a geometry not especially favoured by nickel(II) which is most commonly octahedral in its high spin state.

Larger rings have also been employed for the formation of catenanes. The 56-membered macrocycle **5**, incorporating two 2,9-bis(*p*-biphenylyl)-1,10-phenanthroline fragments, has been demonstrated to form a catenated mixed–ligand complex **6** of type [CuLL^1]$^+$ where L = **5** and L^1 is the smaller ring **3**.[8]

6.2.2   High-yield Synthesis of a [2]-Catenane

A combination of the metal-based template strategy discussed above and a ring-closing metathesis reaction involving intramolecular formation of carbon–carbon

5   A.-M. Albrecht-Gary, Z. Saad, C.O. Dietrich-Buchecker and J.-P. Sauvage, *J. Am. Chem. Soc.*, 1985, **107**, 3205.
6   G. Tartarini, *Gazz. Chim.*, 1933, **63**, 597.
7   C.O. Dietrich-Buchecker, J. Guilhem, J.-M. Kern, C. Pascard and J.-P. Sauvage, *Inorg. Chem.*, 1994, **33**, 3498.
8   J.-M. Kern, J.-P. Sauvage, J.-L. Weidmann, N. Armaroli, L. Flamigni, P. Ceroni and V. Balzani, *Inorg. Chem.*, 1997, **36**, 5329.

5

6

bonds in the presence of ruthenium benzylidene catalyst has been employed to yield a number of [2]-catenane products that include **7** and **8**.[9] The overall reaction scheme

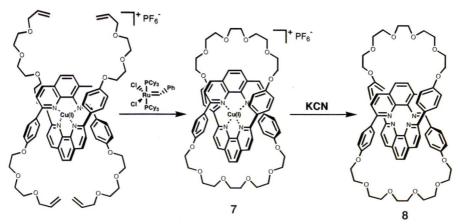

Figure 6.2 *A high-yield synthesis for the [2]-catenane,* **8**[9]

is illustrated in Figure 6.2. Favourable features of the ring-closing metathesis reaction are that it tends to be little affected by the presence of other functional groups and that it proceeds with high efficiency; yields greater than 90% were achieved for the formation of **7** and **8**.

6.2.3 Other Catenanes and Rotaxanes

The successful construction of the single-metal catenane **4** inspired a series of subsequent studies in which catenanes based on more than one metal ion were synthesised.[10–12]

In one example of this type,[10] the [3]-catenane **9** (Figure 6.3) was constructed using an extension of the strategy employed by Sauvage *et al.*[4] in their original synthesis of a [2]-catenane.

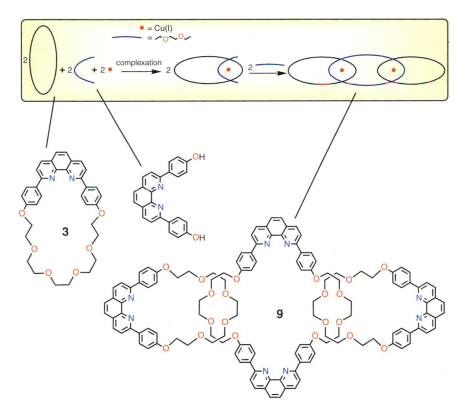

Figure 6.3 *Formation of the [3]-catenane,* **9**[10]

9   B. Mohr, M. Weck, J.-P. Sauvage and R.H. Grubbs, *Angew. Chem., Int. Ed. Engl.*, 1997, **36**, 1308.
10   J.-P. Sauvage and J. Weiss, *J. Am. Chem. Soc.*, 1985, **107**, 6108.
11   C.O. Dietrich-Buchecker, A. Khemiss and J.-P. Sauvage, *J. Chem. Soc., Chem. Commun.*, 1986, 1376.
12   C.O. Dietrich-Buchecker, J. Guilhem, A.K. Khemiss, J.-P. Kintzinger, C. Pascard and J.-P. Sauvage, *Angew. Chem., Int. Ed. Engl.*, 1987, **26**, 661.

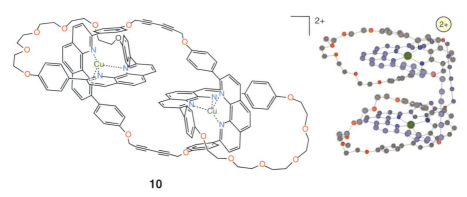

10

Figure 6.4 *A dinuclear copper(I)-containing* [2]-*catenane formed* via *double acetylene oxidative coupling*[12]

In a further example, the final (ring-closing) step involved the double acetylene oxidative coupling of terminal diynes to give the dinuclear copper(I) species **10** in 58% yield.[12] This catenate does not adopt the 'open' arrangement suggested by **10** in Figure 6.4 in either the solid state or in solution; rather, intramolecular forces lead to folding of the large central ring such that its phenanthroline subunits are aligned almost parallel. NMR evidence indicates that a similar folded 'ternary' structure is maintained in solution. In part, the adoption of such an arrangement has been ascribed to the presence of π-interactions between the aromatic rings belonging to each copper complex subunit. Once again, the above product was successfully demetallated by treatment with cyanide ion.[11] The photochemistry and photophysics of variously metallated [copper(I), silver(I) and zinc(II)] products – including heteronuclear systems – were subsequently investigated.[13] In the case of the [3]-catenand (namely, the metal-free system) and its metal complexes (catenates), two or more subunits have been identified that posesses their own excited state properties. Electronic communication between sites in the same compound was shown to lead to partial or complete luminescence quenching.

A wide range of other related catenanes and polycatenanes, many designed with particular functions in mind, have been reported.[14–19] By way of example, a system was constructed to act as a potential model for aspects of the electron-transfer

[13] N. Armaroli, V. Balzani, F. Barigelletti, L. De Cola, L. Flamigni, J.-P. Sauvage and C. Hemmert, *J. Am. Chem. Soc.*, 1994, **116**, 5211.

[14] N. Armaroli, L. De Cola, V. Balzani, J.-P. Sauvage, C.O. Dietrich-Buchecker, J.-M. Kern and A. Bailal, *J. Chem. Soc., Dalton Trans.*, 1993, 3241.

[15] J.-C. Chambron, C.O. Dietrich-Buchecker, J.-F. Nierengarten and J.-P. Sauvage, *J. Chem. Soc., Chem. Commun.*, 1993, 801.

[16] G.-J.M. Gruter, F.J.J. de Kanter, P.R. Markies, T. Nomoto, O.S. Akkerman and F. Bickelhaupt, *J. Am. Chem. Soc.*, 1993, **115**, 12179.

[17] T. Jorgensen, J. Becher, J.-C. Chambron and J.-P. Sauvage, *Tetrahedron Lett.*, 1994, **35**, 4339.

[18] N. Armaroli, V. Balzani, L. De Cola, C. Hemmert and J.-P. Sauvage, *New J. Chem.* 1994, **18**, 775.

[19] J.-L. Weidmann, J.-M. Kern, J.-P. Sauvage, Y. Geerts, D. Muscat and K. Mullen, *Chem. Commun.* 1996, 1243.

process characteristic of the photosynthetic reaction centre.[20] In prior studies by other workers, models for the primary electron-transfer in photosynthesis have frequently involved the use of porphyrin rings linked by a covalent spacer group.[21] In contrast, the above study involved new bis(porphyrin) species in which a [2]-catenane was employed as the spacer; the donor and acceptor porphyrin derivatives being present as pendant groups attached to opposite loops of the catenane. This compound (**11**) was once again constructed around a 'classical' bis(phenanthroline)/copper(I) core of the type employed in the original Sauvage studies. The ¹H NMR and UV–Vis spectra of this product gave no evidence for significant interaction between the two porphyrin rings. However, the luminescence of the zinc(II) porphyrin does undergo significant quenching in the trimetallated species **11**.

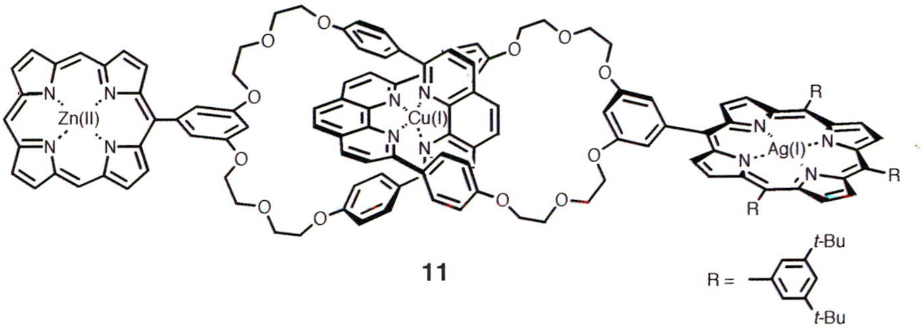

11

Corresponding symmetrical, bis-porphyrin derivatives, as well as larger entities incorporating bis-porphyrin and bis-catenane groups – see **12** – have also been investigated.[22,23]

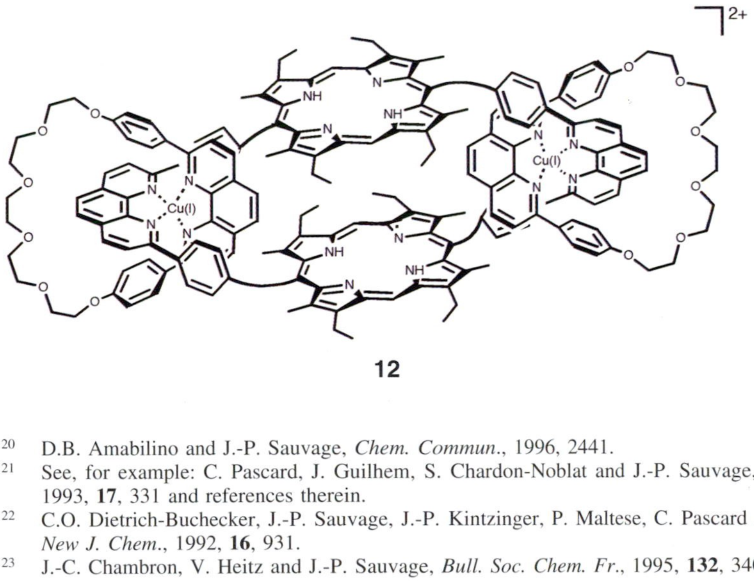

12

20   D.B. Amabilino and J.-P. Sauvage, *Chem. Commun.*, 1996, 2441.
21   See, for example: C. Pascard, J. Guilhem, S. Chardon-Noblat and J.-P. Sauvage, *New J. Chem.*, 1993, **17**, 331 and references therein.
22   C.O. Dietrich-Buchecker, J.-P. Sauvage, J.-P. Kintzinger, P. Maltese, C. Pascard and J. Guilhem, *New J. Chem.*, 1992, **16**, 931.
23   J.-C. Chambron, V. Heitz and J.-P. Sauvage, *Bull. Soc. Chem. Fr.*, 1995, **132**, 340.

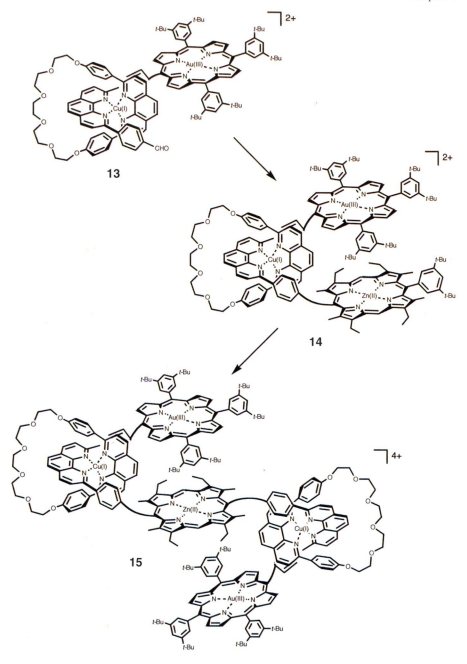

Figure 6.5 *Synthesis of the porphyrin-containing [2]- and [3]-rotaxanes,* **14** *and* **15**[27]

Other studies involving related di- and tri-porphyrin-containing rotaxanes based on linked 1,10-phenanthroline backbones have also been reported.[24–28] One such study involved the initial assembly of the copper(I) complex **13** (Figure 6.5). This precursor is, once again, formed by metal-directed threading of the 2,9-diphenyl-1,10-phenanthroline-containing fragment, incorporating an end-attached gold(III) porphyrin, through the pre-synthesised 30-membered macrocyclic ring. Reaction of **13** with a second porphyrin derivative in a one-pot reaction yielded a mixture of products. These included the target [2]-rotaxane **14** [after metallation with zinc(II) acetate] and an unexpected related [3]-rotaxane **15**, in which a porphyrin moiety links two precursor assemblies.[27] These and other products were separated by chromatography. Treatment of **14** with excess KCN results in generation of the corresponding copper-free [2]-rotaxane.[26]

The electron-transfer properties of rotaxanes of the above type, incorporating Au(III) and Zn(II) porphyrins as the terminal stoppers but with the 'central site' varying through copper(I), zinc(II) or being metal-free, have been investigated.[29] Selective excitation of either porphyrin (using a wavelength at which only the required porphyrin adsorbs) results in rapid electron-transfer from the zinc porphyrin site to the gold porphyrin site. The ground state is then restored by a relatively slow reverse electron-transfer reaction. Not unexpectedly, the nature of the 'spacer' moiety (and the corresponding structure adopted) has a marked effect on the rates of the electron-transfer processes occurring within these bis-porphyrin-stoppered rotaxanes. For the case where copper(I) is present in the 'central' complex, the effect of this ion in mediating the photoinduced electron-transfer between the porphyrin subunits has been studied.[30] The copper(I) complex is postulated to donate an electron to the zinc porphyrin radical cation initially generated, with the ground state system being restored by relatively slow electron-transfer from the gold porphyrin neutral radical to the resulting copper(II) complex.

A variation of the above strategy has been used to produce the related [2]-rotaxane **16** in 15% yield. In this case, two potentially redox-active fullerenes act as the stoppers.[31] C_{60} was chosen for inclusion in this product because of its interesting electrochemical and electronic properties and, in particular, because it is a strong electron acceptor. Although the substituted fullerene stoppers appear to have a significant influence on the redox properties of the metal centre (an anodic shift occurs), the converse is not true and the metal does not significantly affect the redox properties of the fullerene groups.

24   J.-C. Chambron, S. Chardon-Noblat, A. Harriman, V. Heitz and J.-P. Sauvage, *Pure Appl. Chem.*, 1993, **65**, 2343.
25   M. Linke, J.-C. Chambron, V. Heitz and J.-P. Sauvage, *J. Am. Chem. Soc.*, 1997, **119**, 11329.
26   J.-C. Chambron, V. Heitz and J.-P. Sauvage, *J. Am. Chem. Soc.*, 1993, **115**, 12378.
27   J.-C. Chambron, C.O. Dietrich-Buchecker, V. Heitz, J.-F. Nierengarten, J.-P. Sauvage, C. Pascard and J. Guilhem, *Pure Appl. Chem.*, 1995, **67**, 233.
28   N. Solladie, J.-C. Chambron, C. Dietrich-Buchecker and J.-P. Sauvage, *Angew. Chem., Int. Ed. Engl.*, 1996, **35**, 906.
29   J.C. Chambron, A. Harriman, V. Heitz and J.-P. Sauvage, *J. Am. Chem. Soc.*, 1993, **115**, 7419.
30   J.-C. Chambron, A. Harriman, V. Heitz and J.-P. Sauvage, *J. Am. Chem. Soc.*, 1993, **115**, 6109.
31   F. Diederich, C. Dietrich-Buchecker, J.-F. Nierengarten and J.-P. Sauvage, *J. Chem. Soc., Chem. Commun.* 1995, 781.

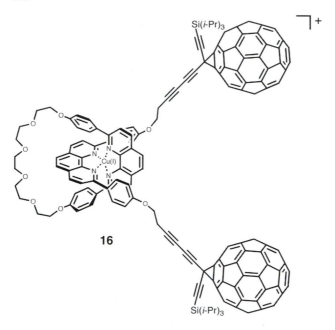

16

As might be anticipated, the nature of the product formed in the metal-directed threading of molecular strings into a co-ordinating macrocyclic ligand has been demonstrated to be influenced by the type of the metal-binding site present, the ratio of the individual components employed and the rigidity/flexibility of the strings to be threaded.[15]

A number of interlocking and threaded rings have been constructed based on the preferential affinity of bis-1,10-phenanthroline fragments for copper(I) as well as of bis-terpyridine fragments for ruthenium(II).[32,33]

In one such study, the key organic precursor **17** (Figure 6.6) incorporates two terpyridine derivative centres and one 2,9-diphenyl-1,10-phenanthroline centre. Interaction of this compound with macrocycle **3** in the presence of copper(I) leads to its copper-directed threading through the phenanthroline-containing ring to yield **18**. Co-ordination of ruthenium(II) at the terminal terpyridyl fragments then yields the corresponding mixed-metal [2]-rotaxane **19** and the [2]-catenane **20** without the need for further organic reaction. The high stability of the co-ordinated ruthenium centres has allowed the selective removal of copper from each of these species (by treatment with cyanide ion in MeCN–H$_2$O). These preparations thus represent relatively facile procedures for generating new multi-component systems incorporating potentially 'useful' ruthenium(II) photoactive centres.

[32] D.J. Cardenas, P. Gavina and J.-P. Sauvage, *Chem. Commun.*, 1996, 1915.
[33] D.J. Cardenas, P. Gavina and J.-P. Sauvage, *J. Am. Chem. Soc.*, 1997, **119**, 2656.

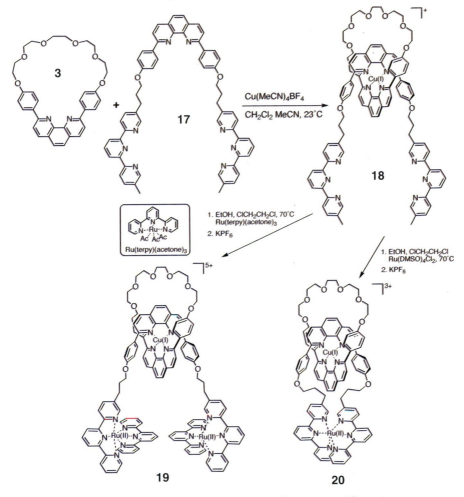

Figure 6.6 *Synthesis of a mixed copper(I)/ruthenium(II) rotaxane,* **19**, *and catenane,* **20**[32,33]

6.2.4 Poly-metallorotaxanes Containing Conjugated Rods

Related poly-metallorotaxanes containing conjugated rods threaded through co-ordinating macrocyclic rings incorporating the usual 1,10-phenanthroline fragment(s) have been reported.[34–37] For example, 'rigid rack' di- and tri-copper(I) rotaxanes of type **21–23** (Figure 6.7) readily self-assemble in solution when the

34   S.S. Zhu, P.J. Carroll and T.M. Swager, *J. Am. Chem. Soc.*, 1996, **118**, 8713.
35   P.N.W. Baxter, H. Sleiman, J.-M. Lehn and K. Rissanen, *Angew. Chem., Int. Ed. Engl.*, 1997, **36**, 1294.
36   H. Sleiman, P.N.W. Baxter, J.-M. Lehn, K. Airola and K. Rissanen, *Inorg. Chem.*, 1997, **36**, 4734.
37   P.L. Vidal, M. Billon, B. Divisia-Blohorn, G. Bidan, J.-M. Kern and J.-P. Sauvage, *Chem. Commun.*, 1998, 629.

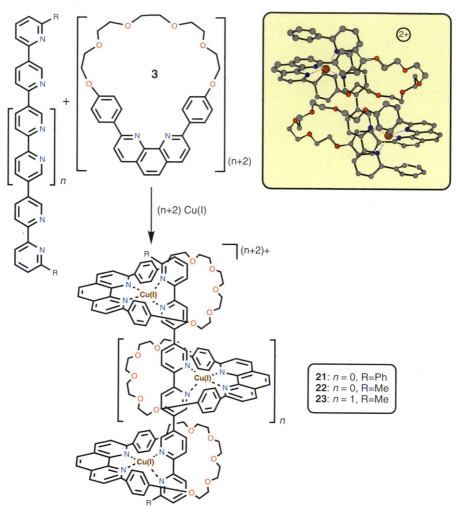

Figure 6.7 *Formation of polyrotaxanes. Insert shows the structure of the dinuclear copper(I) species for the system with* n = O, R = Ph[38]

appropriate polypyridine rod and the phenanthroline-containing macrocycle are mixed in the presence of $[Cu(MeCN)_4]^+$.[38] The structure of the dinuclear derivative (**21** ; R = Ph, *n* = 0) has been confirmed by X-ray diffraction (Figure 6.7).

In another study of this type, a polyrotaxane incorporating a phenanthroline/thiophene-based conjugated backbone was synthesised *via* a copper(I) template reaction, followed by electropolymerisation. After removal of the copper(I), re-metallation was only found to be possible if lithium ion had been present during the original copper decomplexation reaction. The lithium ion appears to act

38   H. Sleiman, P. Baxter, J.-M. Lehn and K. Rissanen, *J. Chem. Soc., Chem. Commun.*, 1995, 715.

as a labile scaffolding that maintains the topology of the co-ordination sites after copper removal.

6.2.5 Further Polynuclear Systems

The three-metal catenane **24**, incorporating one copper(I) and two ruthenium(II) ions, has been synthesised by means of the copper(I) template reaction illustrated in Figure 6.8.[39] The double clipping procedure was found to be selective for the

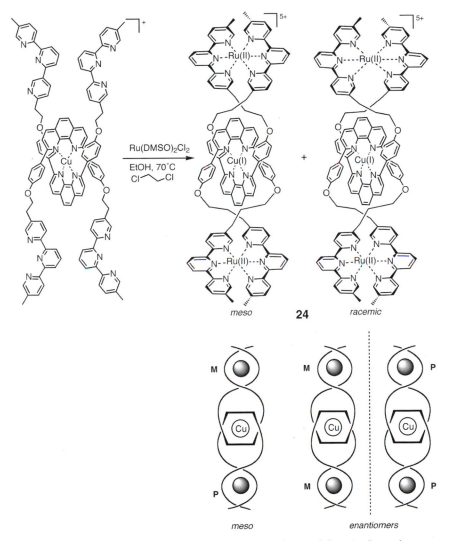

Figure 6.8 *Template synthesis of the meso and racemic forms of the mixed metal helicate,* **24**[39]

39   D.J. Cardenas and J.-P. Sauvage, *Inorg. Chem.*, 1997, **36**, 2777.

formation of the catenated product over other topologies; the product was formed as an equimolar mixture of diastereoisomers (a racemic mixture and a meso form). Other heterometallic copper(I)-containing species of the above general type, incorporating iron(II), cobalt(II), copper(II) or zinc(II), were also synthesised in this study.

In an extension of the general synthetic strategies discussed so far, the Sauvage group has been successful in synthesising other multi-ring catenanes and rotaxanes, albeit in small yield.[40–42] One such study produced a novel family of interlocked systems which were characterised by electrospray mass spectrometry. These products contain between two and six smaller rings threaded onto a much larger ring (Figure 6.9).[40]

In another study, success was achieved in 'clipping' a phenanthroline–polyether ring of the usual type on to each of three arms of a much larger cryptand-like skeleton that incorporated a phenantroline group in each arm (Figure 6.10).[41] The prod-

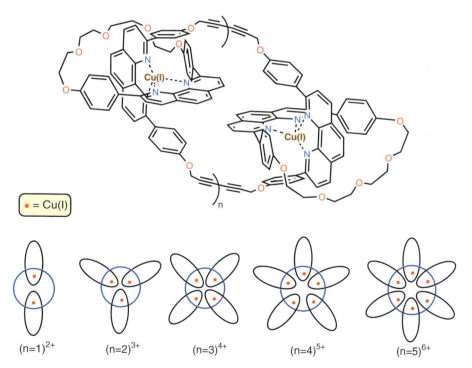

Figure 6.9 *A family of novel interlocked rings formed in low yield and characterised by electrospray mass spectrometry[40]*

[40]   F. Bitsch, C.O. Dietrich-Buchecker, A.-K. Khemiss, J.-P. Sauvage and A.V. Dorsselaer, *J. Am. Chem. Soc.,* 1991, **113**, 4023.
[41]   C. Dietrich-Buchecker, B. Frommberger, I. Luer, J.-P. Sauvage and F.Vögtle, *Angew. Chem., Int. Ed. Engl.,* 1993, **32**, 1434.
[42]   F. Bitsch, G. Hegy, C.O. Dietrich-Buchecker, E. Leize, J.-P. Sauvage and A. Van Dorsselaer, *New J. Chem.,* 1994, **18**, 801.

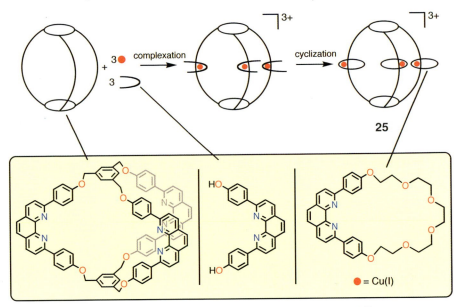

Figure 6.10 *Low yield synthesis of the novel 'cage-like' tricatenane,* **25**[41]

uct was of type **25**. However, as expected, the yield in this case was miniscule at half of one percent. Very small yields of the corresponding one- and two-linked catenanes were also isolated from the reaction mixture in this case.

6.2.6 Redox Switching

Molecular systems whose configuration can be altered by means of an external stimulus are of much interest as potential molecular switches for information storage. An electrochemically-induced molecular rearrangement within a copper-containing catenane of the type illustrated in Figure 6.11 has been reported.[43–45] The control of this simple 'molecular device' involves a switching of the oxidation state (between +1 and +2) of the copper ion co-ordinated to both interlinked rings of the catenane; such redox switching can also be achieved chemically or photochemically for this system. The 'gliding' of one ring within the other about the central copper is reflected by a configurational change that enables the central metal ion to achieve its preferred co-ordination geometry for each oxidation state: four (N_4-tetrahedral) for copper(I), five (N_5-square pyramidal) for copper(II) in this system. Accompanying the co-ordination change is an alteration of the electronic configuration of the copper: from d^{10} for copper(I) to d^9 for copper(II), with the latter being paramagnetic and detectable by EPR spectroscopy. The nature of the EPR spectrum was used to infer the presence of square pyramidal co-ordination in the copper(II) case.

[43]   F. Baumann, A. Livoreil, W. Kaim and J.-P. Sauvage, *Chem. Commun.*, 1997, 35.
[44]   C.O. Dietrich-Buchecker, A. Livoreil and J.-P. Sauvage, *J. Am. Chem. Soc.*, 1994, **116**, 9399.
[45]   A. Livoreil, J.-P. Sauvage, N. Armaroli, V. Balzani, L. Flamigni and B.J. Ventura, *J. Am. Chem. Soc.*, 1997, **119**, 12114.

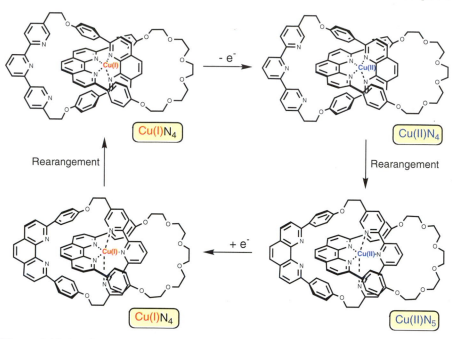

Figure 6.11 *An electrochemically-induced molecular rearrangement of a copper(I)-containing 'unsymmetrical' [2]-catenane*[43-45]

In a related set of experiments,[46] redox control of the ring-gliding motion in the copper-complexed catenane, derived from two identical macrocyclic rings containing both bipyridine and terpyridine metal binding sites, has been investigated. For this system, electrochemical control of molecular motion between *three* distinct co-ordination geometries was documented. Each step resulted in a different co-ordination number of the central copper ion; corresponding to the respective co-ordination of four, five and six heterocyclic nitrogen donors to copper. The process is illustrated in Figure 6.12. Beginning with the tetra-co-ordinate copper(I) ion surrounded by two phenanthroline fragments (one from each macrocyclic ring), oxidation leads to the formation of an intermediate copper(II) complex in which the copper is surrounded by both a phenanthroline and a terpyridyl subunit (again belonging to different macrocyclic rings). This copper(II) species then interconverts to a more stable, hexa-co-ordinate copper(II) complex *via* further ring rotation to yield an N_6-co-ordination sphere made up of nitrogen donor atoms from two terpyridyl units. Reduction then reverses the process.

A similar concept has been applied to the switching of a macrocyclic component **3** between two different co-ordination sites on the linear component **26** to yield the corresponding pseudo-rotaxane.[47] A subsequent report describes parallel studies involving an analogous [2]-rotaxane (incorporating bulky terminal mesylate

[46]   D.J. Cardenas, A. Livoreil and J.-P. Sauvage, *J. Am. Chem. Soc.*, 1996, **118**, 11980.
[47]   J.-P. Collin, P. Gavina and J.-P. Sauvage, *Chem. Commun.*, 1996, 2005.

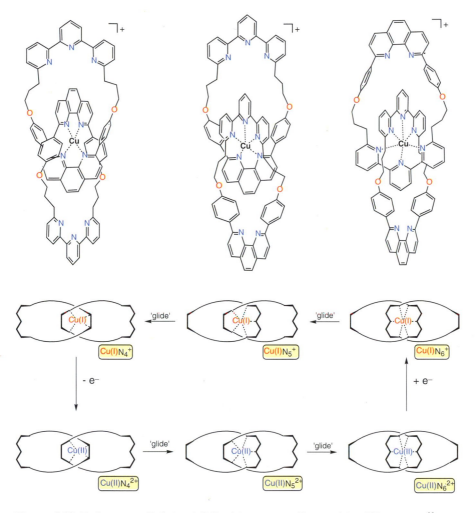

Figure 6.12 *Redox-controlled ring 'gliding' in a copper(I)-containing [2]-catenane*[46]

stoppers).[48] In each system, alternate reduction and oxidation of the copper centre results in the cyclic sequence shown in Figure 6.13; the difference in the preferred co-ordination numbers of copper(I) (four) and copper(II) (five) results in 'hopping' of the copper-containing cyclic component between the respective phenanthroline and terpyridine sites on the linear component.

6.2.7 A Hybrid [2]-Catenane

A hybrid 'bimodal' [2]-catenane incorporating structural elements from the catenane types developed by both the Stoddart and Sauvage groups in the one system

---

[48]   P. Gaviá and J.-P. Sauvage, *Tetrahedron Lett.*, 1997, **38**, 3521.

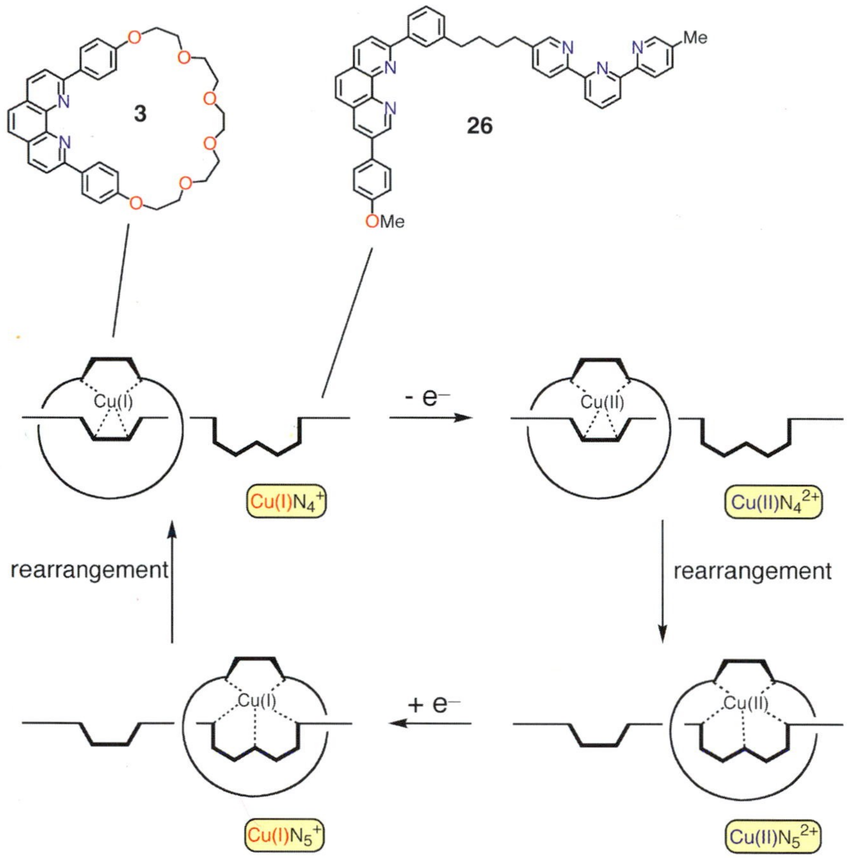

Figure 6.13 *A pseudo-rotaxane capable of switching between two arrangements under redox control*[47]

has been prepared. Details of the structure are given by **28** (and **27**) in Figure 6.14. Once again, a copper(I) template procedure formed a central part of the overall synthesis.[49] The resulting novel system incorporates a metal-ion binding site (1,10-phenanthroline fragment) together with a π-electron-rich entity (1,5-dioxynaphthalene fragment) in one loop of the catenane while the second loop contains a further 1,10-phenanthroline residue and two π-electron-deficient groups (bipyridinium units) of the type used previously for formation of donor–acceptor complexes (see Chapters 3 and 4). Once again, the rings can adopt two favoured orientations with respect to each other simply by rotation of one ring within the other. This major topological rearrangement of the molecule may be triggered by removing or adding a cation (Cu+, Li+ or H+) at the bis-phenanthroline complexation site. In the metal binding mode, the two phenanthroline units adopt the usual tetrahedral arrangement

49 D.B. Amabilino, C.O. Dietrich-Buchecker, A. Livoreil, L. Perez-Garcia, J.-P. Sauvage and J.F. Stoddart, *J. Am. Chem. Soc.*, 1996, **118**, 3905.

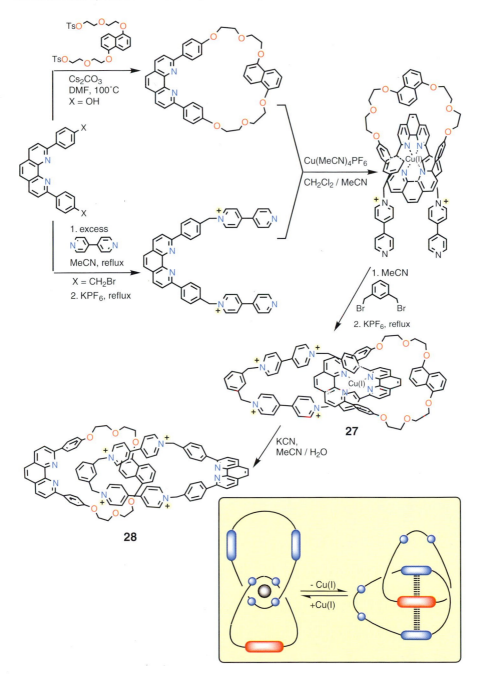

Figure 6.14 *Synthesis of the cationic hybrid [2]-catenane, 28*[49]

about the metal ion. In the metal-free mode, the π-electron-rich and π-electron-deficient units stack to yield a donor–acceptor complex of the type well documented in the previous studies (see insert to Figure 6.14).

6.3 Helicates

Since the structure of DNA was first revealed by Watson and Crick in 1953,[50] the formation of the double helix of nucleic acids by a self-directing, co-operative process has fascinated several generations of scientists. In general terms, DNA may be seen to self-assemble by a process in which two long chain polytopic receptors interact with the base pairs in such a way that each association step sets the stage for the one that follows.

The formation of three-stranded, helical (synthetic) proteins has been shown to be influenced by metal-ion binding to ligands that form part of the component peptide strands. In one study of this type,[51] a 15-residue amphiphilic peptide containing a 2,2′-bipyridine derivative situated at the N-terminus was demonstrated to self-assemble spontaneously in the presence of selected transition metal ions to form a 45-residue metalloprotein with a triple-helical, 'coiled-coil' structure.

The development of metal template strategies for mimicking Nature in the spontaneous assembly of totally synthetic helical structures represents an important milestone in the development of supramolecular chemistry.[52–55] Athough chirality in classical octahedral metal complexes has been both recognised and investigated since Alfred Werner's time, it is only relatively recently that considerable interest has been generated in synthesising di- and oligo-metallic species exhibiting chiral helical structures. In these syntheses the binding of one metal ion can be thought of as 'setting up' the system for the binding of the next such that the reaction proceeds in a progressive manner towards the final helical product – provided, of course, that the latter is the most stable of all the possible products.

6.4 Single-stranded Helicates

Single-stranded helical co-ordination around a single metal ion was reported many years ago,[56] and other examples of complexes of this type have continued to appear[57,58] together with reports of single-stranded dimetallic[59] and oligometallic[60,61] systems.

50   J.D. Watson and F.H.C. Crick, *Nature*, 1953, **171**, 737.
51   M.R. Ghadiri, C. Soares and C. Choi, *J. Am. Chem. Soc.,* 1992, **92**, 825.
52   E.C. Constable, *Tetrahedron*, 1992, **48**, 10013.
53   J.-C. Chambron, C. Dietrich-Buchecker and J.-P. Sauvage, *Topics Current Chem.*, 1993, **165**, 131.
54   E.C. Constable, *Prog. Inorg. Chem.*, 1994, **42**, 67.
55   C. Piguet, G. Bernardinelli and G. Hopfgartner, *Chem. Rev.*, 1997, **97**, 2005.
56   L.F. Lindoy and D.H. Busch, *Inorg. Chem.*, 1974, **13**, 2494.
57   M.L. Tulchinsky, L.I. Demina, S.V. Lindeman and Y.T. Struchkov, *Inorg. Chem.*, 1994, **33**, 5836 and references therein.
58   J. Christoffers and R.G. Bergman, *Angew. Chem., Int. Ed. Engl.*, 1995, **34**, 2266.
59   C.J. Cathey, E.C. Constable, M.J. Hannon, D.A. Tocher and M.D. Ward, *J. Chem. Soc., Chem. Commun.*, 1990, 621.
60   O.J. Gelling, F. Van Bolhuis and B.L. Feringa, *J. Chem. Soc., Chem. Commun.*, 1991, 917.
61   P.K. Bowyer, K.A. Porter, A.D. Rae, A.C. Willis and S.B. Wild, *Chem. Commun.*, 1998, 1153.

6.5 Double-stranded Helicates

Reports of double-helical complexes have appeared in the literature since the sixties.[62] Despite the early interest, it is only more recently that emphasis has been given to the use of metal template synthesis for obtaining a wide range of doubly- and triply-stranded systems. In part, this attention has had its origins in an early report by Lehn *et al.*[63] in which the spontaneous assembly of a dicopper(I)- containing double helix was described.

Like Lehn's seminal investigation, which is discussed later in this section, a con- siderable number of subsequent studies have involved semi-rigid 'oligo-polyden- tate' ligands. Such ligands have been employed in the metal-ion-directed self-assembly of double- and triple-helical complexes containing metal ions that are positioned along the helical axis in the final structures. Key to the formation of such an arrangement is the need for the stereochemical preferences of both the metal ions and the ligands to be met and, for example, that the selection between mono-, double- and triple-helical structures can be greatly influenced by the size and co-ordination preferences of the chosen metal ion. Thus, a metal ion with a preference for four (tetrahedral) co-ordination will tend to interact with a semi- rigid, potentially bis-bidentate ligand to yield a double helix of stoichiometry $[M_2L_2]^{n+}$. On the other hand, if the metal ion shows a preference for octahedral co- ordination then a triple-helical species of stoichiometry $[M_2L_3]^{n+}$ may form.[64] Triple- helical structures are discussed later in this chapter.

6.5.1 Systems Based on Di- and Oligo-bipyridyl Derivatives

The Lehn group synthesised the trinuclear copper(I) complex of the substituted 2,2'- bipyridyl/ether derivative **29** and confirmed its helical structure by X-ray diffrac- tion (Figure 6.15).[65] A dinuclear copper(I) complex of **30** (Figure 6.15) was also synthesised and subsequently the corresponding 'dimeric' ethane-bridged bipyri- dine and phenanthroline ligands were also demonstrated to yield similar di-helicate copper(I) species. These results thus indicate that helicate formation is tolerant towards minor modification of ligand structure.[66]

In related studies, the syntheses of the corresponding tetra- and penta-nuclear copper(I) helicates, based on ligands respectively incorporating four and five

[62]   D. Dolphin, R.L.N. Harris, J.L. Huppatz, A.W. Johnson, I.T. Kay and J. Leng, *J. Chem. Soc., C,* 1966, 98; G. Struckmeier, U. Thewalt and J.-H. Fuhrhop, *J. Am. Chem. Soc.,* 1976, **98**, 278; D. Wester and G.J. Palenik, *Inorg. Chem.,* 1976, **15**, 755; W.S. Sheldrick and J. Engel, *Acta Crystallogr. Sect. B,* 1981, **37**, 250; G.C. van Stein, G. van Koten, F. Blank, L.C. Taylor, K. Vrieze, A.L. Spek, A.J.M. Duisenberg, A.M.M. Schreurs, B. Kojic-Prodic and C. Brevard, *Inorg. Chim. Acta,* 1985, **98**, 107; G.C. van Stein, G. van Koten, K. Vrieze, A.L. Spek, E.A. Klop and C. Brevard, *Inorg. Chem.,* 1985, **24**, 1367 and references therein.
[63]   J.-M. Lehn, J.-P. Sauvage, J. Simon, R. Ziessel, C.P. Leopardi, G. Germain, J.-P. Declercq and M. Van Meersche, *New J. Chem.,* 1983, **7**, 413.
[64]   A.F. Williams, *Pure Appl. Chem.,* 1996, **68**, 1285.
[65]   J.-M. Lehn, A. Rigault, J. Siegel, J. Harrowfield, B. Chevrier and D. Moras, *Proc. Natl. Acad. Sci. USA,* 1987, **84**, 2565.
[66]   M.-T. Youinou, R. Ziessel, and J.-M. Lehn, *Inorg. Chem.,* 1991, **30**, 2144.

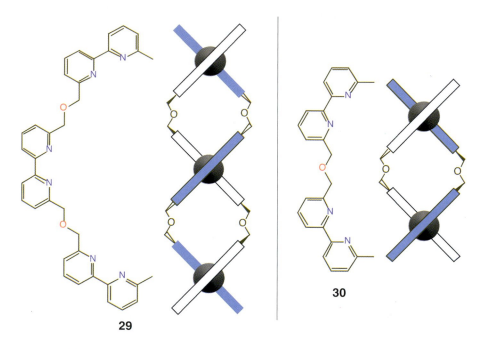

29

30

Figure 6.15 *The helical arrangements in the copper(I) complexes of* **29** *and* **30**[65]

bipyridyls in each strand, were achieved.[67,68] These deep red copper(I) products were estimated to have overall unstrained lengths of about 22 and 27 Å, respectively; thus bringing them into the realm of nanostructures.

Reaction of the oligo-bipyridyl ligands[68] of the type just discussed containing three, four and five bipyridyl subunits with silver(I) trifluoromethanesulfonate in acetonitrile results in assembly of the corresponding trinuclear, tetranuclear and pentanuclear silver(I) helicates; the double-helical structure of the trinuclear species has been confirmed by X-ray diffraction.[69] Clearly, the silver(I) ion is also capable of promoting efficient helical self-assembly, in spite of the fact that its binding strength to bipyridyl groups is known to be significantly lower than occurs for copper(I).

Equilibrium studies involving double-stranded trihelicates have demonstrated the co-operative nature of metal ion binding. A spectrophotometric titration investigation of the stepwise binding of copper(I) to a (substituted) tris-bipyridyl ligand system of the present type in dichloromethane–acetonitrile (1 : 1) indicated the formation of a single, well-defined product.[70] From analysis of the data, the self-

[67] J.-M. Lehn and A. Rigault, *Angew. Chem., Int. Ed. Engl.,* 1988, **27**, 1095.
[68] M.M. Harding, U. Koert, J.-M. Lehn, A. Marquis-Rigault, C. Piguet and J. Siegel, *Helv. Chim. Acta,* 1991, **74**, 594.
[69] T.M. Garrett, U. Koert, J.-M. Lehn, A. Rigault, D. Meyer and J. Fischer, *J. Chem. Soc., Chem. Commun.,* 1990, 557.
[70] A. Pfeil and J.-M. Lehn, *J. Chem. Soc., Chem. Commun.,* 1992, 838.

assembly of the resulting helicate was seen to display positive co-operativity; namely, helicate formation is driven towards completion by successive metal-ion binding.

A similar result was obtained for the corresponding silver(I) trihelicates using both spectrophotometric and potentiometric (silver electrode) data.[71] Thus, the stability constant K_2 [corresponding to reaction (6.2)] is greater than K_1 [for reaction (6.1)]. In fact, for positive co-operativity, it is sufficient that $K_2 > (K_1/3)$.

$$[AgL_2]^+ + Ag^+ \rightleftharpoons [Ag_2L_2]^{2+} \tag{6.1}$$

$$[Ag_2L_2]^{2+} + Ag^+ \rightleftharpoons [Ag_3L_2]^{3+} \tag{6.2}$$

It was suggested that, on initial complexation of a metal by the two ligand strands, the stability constants for the binding of subsequent metal ions may be enhanced somewhat by the occurrence of attractive π-stacking interactions between bipyridyl groups as the structure assemblies (although the magnitude of such a contribution is uncertain).

Lehn *et al.* have used oligo 2,2'-bipyridyl strands to illustrate a further type of recognition – the recognition of *like by like*.[72] Reaction of copper(I) ions with a mixture of oligo 2,2'-bipyridyl strands containing from two to five bipyridyl units gives an impressive example of recognition of this type. The double helicates formed were demonstrated to contain, almost exclusively, matching ligand strands. That is, no significant formation of mixed strand complexes occurred. In part, this may be seen as reflecting the maximisation of donor atom–metal ion bond formation across the total metal–mixed ligand system.

Deoxyribonucleo-functionalised oligo-bipyridyl ligands and their copper(I) complexes, such as the pentahelicate **31**, have been synthesised. In these, the substituents are orientated on the periphery of the double helix such that they point outwards.[73] In contrast to DNA, these helical structures thus have positive charges located inside the strands of the helix while the nucleic acid bases are on the outside. Although the presence of the chiral nucleosides could lead to preferential induction of one helical sense, whether this is the case or not was not determined in this study.

Further 2,2'-bipyridine derivatives incorporating attached nucleoside and amino acid recognition units have been prepared.[74] The latter groups appear suitable for *exo*-metal co-ordination to a second metal as well as for recognition of biological target molecules (such as nucleic acids) through both direct electrostatic and hydrogen bond interactions.

The interaction of double-helical copper(I) species incorporating a selection of the dimethyl-substituted, oxybis(methylene)-linked oligo-bipyridines[67] discussed earlier with double-stranded DNA has been investigated.[75] Amongst other effects,

[71] T.M. Garrett, U. Koert and J.-M. Lehn, *J. Phys. Org. Chem.*, 1992, **5**, 529.

[72] R. Krämer, J.-M. Lehn and A. Marquis-Rigault, *Proc. Natl. Acad. Sci. USA*, 1993, **90**, 5394.

[73] U. Koert, M.M. Harding and J.-M. Lehn, *Nature*, 1990, **346**, 339.

[74] M.M. Harding and J.-M. Lehn, *Aust. J. Chem.*, 1996, **49**, 1023.

[75] B. Schoentjes and J.-M. Lehn, *Helv. Chim. Acta*, 1995, **78**, 1.

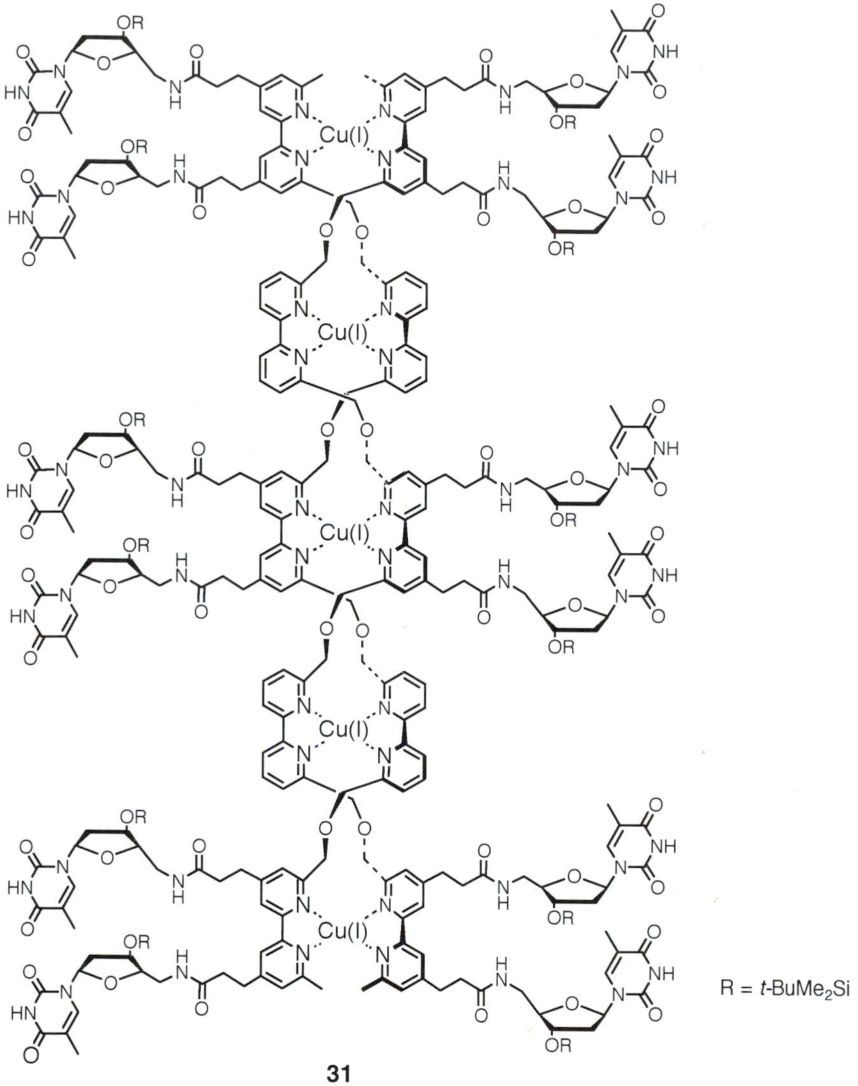

R = *t*-BuMe₂Si

31

the helicates were shown to inhibit the cleavage of DNA by two restriction enzymes – behaviour that is in accord with binding in the major groove of DNA having occurred.

In an impressive demonstration of self-assembly, Lehn and co-workers[76] showed that **32** (Figure 6.16) reacts with an equimolar amount of ferric chloride in ethylene glycol at 170 °C to yield the red solution expected for the presence of an iron(II) tris-bipyridyl chromophore. Molecular models indicated that the three bipyridyl fragments of each ligand are best accommodated by co-ordination to three differ-

76 B. Hasenknopf, J.-M. Lehn, B.O. Kneisel, G. Baum and D. Fenske, *Angew. Chem., Int. Ed. Engl.*, 1996, **35**, 1838.

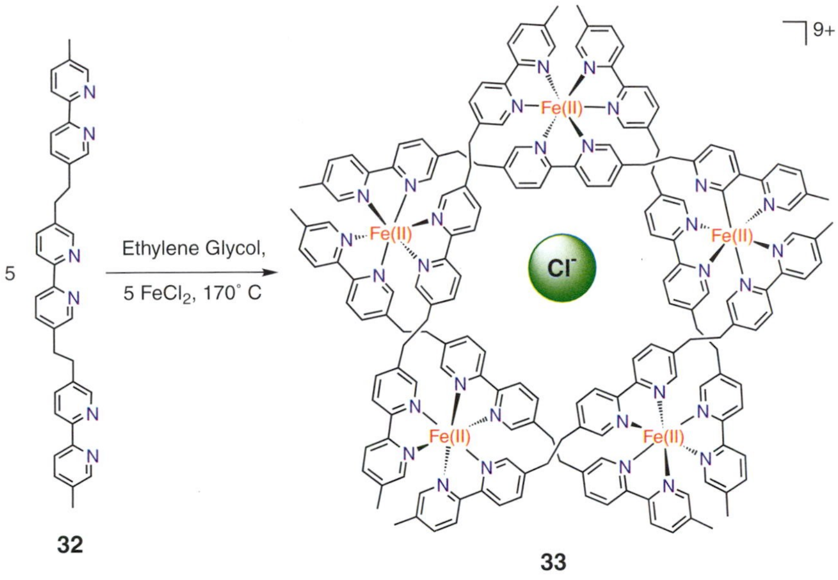

Figure 6.16 *Circular, double-helical structure of $[(Fe_5L_5)\cdot Cl]^{9+}$ $(L = 32)$, 33*[76]

ent iron(II) atoms. Addition of hexafluorophosphate anion to the red solution led to isolation of $[(Fe_5L_5)Cl](PF_6)_9$ as a red solid; the electrospray mass spectrum of this compound confirmed its stoichiometry. The chloride ion was assumed to be strongly bound since it was present in all spectral fragments. Finally, the X-ray structure of the product **33** confirmed its circular, double-helical structure. The aesthetically pleasing structure exhibits five-fold symmetry and contains the chloride ion bound in the inner cavity of the torus. The overall diameter of the torus is about 22 Å. In an extension of this study the self-assembly of further tetra- and hexa-nuclear, circular helicates has been investigated.[77] There is evidence that the diameter of the chloride ion influences the size of the torus observed in this case.

The self-assembly of a new category of tetranuclear complexes of both copper(I) and silver(I) has been reported.[78] Interaction of these metal ions with the (bis)bipyridine-based macrocycle **34** (Figure 6.17), incorporating phenanthroline pendant arms, results in the assembly of 4 : 2 (metal : ligand) complexes incorporating four helical domains. An X-ray diffraction study of the species of type $[Cu_4L_2]^{4+}$ showed that it has the topology illustrated in Figure 6.17. Each copper(I) has a distorted tetrahedral geometry. The four co-planar copper(I) ions form a rhombus, with the four positive charges on the cation being balanced by unco-ordinated perchlorate ions. Two of the copper(I) ions fall on the imposed C_2 axis. A similar double-helical structure was also proposed for the analogous silver complex.

[77] B. Hasenknopf, J.-M. Lehn, N. Boumediene, A. Dupont-Gervais, A. Van Dorsselaer, B. Kneisel and D. Fenske, *J. Am. Chem. Soc.*, 1997, **119**, 10956.
[78] R. Ziessel and M.-T. Youinou, *Angew. Chem., Int. Ed. Engl.*, 1993, **32**, 877.

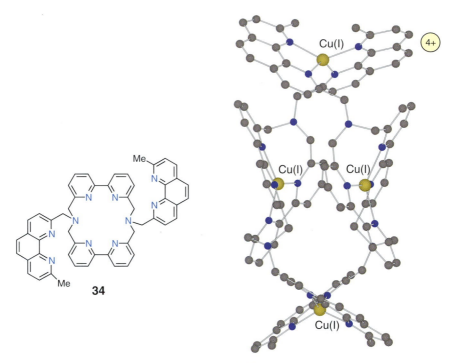

Figure 6.17 *Unusual helical structure of $[Cu_4L_2]^{4+}$ (L = 34) in which the four copper(I) ions are coplanar, with each copper being tetrahedrally co-ordinated*[78]

6.5.2 Systems Based on Directly-linked, Oligo-pyridines

In an extensive series of studies, Constable *et al.*[52,54] investigated the complexation of open-chain, 2,6-linked oligopyridines incorporating between four and six pyridyl rings. These conjugated ligands do not incorporate spacer groups between the respective heterocyclic rings and hence tend to assume moderately rigid structures. A strictly co-planar arrangement of the type suitable for co-ordination to a single metal ion is expected to be destabilised to some degree by steric interactions between the protons situated *ortho* to the respective linked atom positions. Further, the restricted flexibility of this system can result in the orientation of all nitrogen lone pairs not being ideal for co-ordination to a single metal ion – especially when the metal ion is of small radius. This, coupled with the possibility that (ideally) a 180° twist may occur between any pair of adjacent rings, means that these systems have inherent attributes that will favour only a restricted number of alternative co-ordination modes. The interesting metal ion chemistry observed for this ligand series appears to be a direct consequence of the presence of such 'programmed' electronic and steric features. For example, the formation of helicates is very frequently associated with the 'partitioning' of the ligand's donor atom set into two separate metal binding domains through rotation about an interannular C–C bond.

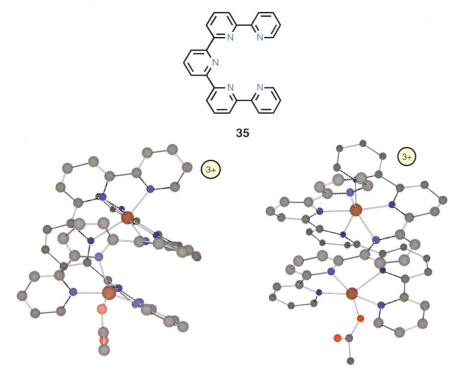

35

Figure 6.18 *Two views of the crystal structure of [Cu₂L₂(OAc)]³⁺ (L = **35**)*[79]

The quinquepyridine ligand **35** (Figure 6.18) interacts with cobalt(II), nickel(II) copper(II) and zinc(II) acetate in boiling methanol[79] to yield 2 : 2 (M : L) cationic complexes that were isolated as their hexafluorophosphate salts. The crystal structure of one product, $[Cu_2L_2(OAc)][PF_6]_3 \cdot H_2O$ (Figure 6.18), confirms the double-helical arrangement of the ligand strands around the copper centres. The copper(II) ions lie 4.50 Å apart, and each quinquepyridine ligand acts as a bidentate to one copper but as a tridentate to the other. An acetate ligand completes the co-ordination sphere of the 'less-bound' copper site. Significant π-stacking appears to be present between planar regions on each ligand; this is reminiscent of the base-pair stacking characteristic of nucleic acids. X-Ray diffraction studies have also confirmed the double-helical nature of the corresponding dinuclear nickel(II)[80] and cobalt(II)[81] complexes. In contrast, **35** forms a shallow, single-helical complex with silver(I); however, the X-ray structure of the diprotonated form of the free ligand (as its tetrafluorophosphate salt) reveals that this cation is near planar – indicating that charge alone is not sufficient to cause helication in this case.[82]

[79] E.C. Constable, M.G.B. Drew and M.D. Ward, *J. Chem. Soc., Chem. Commun.*, 1987, 1600.
[80] E.C. Constable, M.D. Ward, M.G.B. Drew and G.A. Forsyth, *Polyhedron*, 1989, **8**, 2551.
[81] E.C. Constable, S.M. Elder, P.R. Raithby and M.D. Ward, *Polyhedron*, 1991, **10**, 1395.
[82] E.C. Constable, S.M. Elder, J.V. Walker, P.D. Wood and D.A. Tocher, *J. Chem. Soc., Chem. Commun.*, 1992, 229.

   As an aside, it is noted that the latter may not be a hard and fast rule, at least when other ligand types are present. Sauvage, Balzani *et al.*[83] used NMR data as well as absorption and fluorescence results to demonstrate the protonation-driven formation of a double-helical structure involving two ligand strands derived from a pair of 2-*p*-anisyl-1,10-phenanthroline units linked in their 9-positions by a 1,3-phenylene spacer. In this case, the first protonation step does not simply represent mono-protonation of one of the phenanthroline units but rather appears to involve a co-operative interaction with a pair phenanthroline units (on different strands).

   In other studies involving poly-pyridyl ligands, Che *et al.*[84] have shown that quin-quepyridine **35** also yields double-helical ruthenium complexes. The double-helical structure of the $[Ru_2L_2(C_2O_4)]^{2+}$ cation, incorporating a 'terminal' oxalato group, is shown in Figure 6.19. These authors point out that two classes of ruthenium-polypyridyl complexes have been widely studied in recent times. The first are the ruthenium(II) complexes incorporating heterocyclic diamines {of which [Ru(bipyri-dine)$_3$]$^{2+}$ is typical}; these have been used widely as photo-induced electron-transfer reagents. The second category consists of high-valent, oxo-ruthenium complexes; the redox chemistry of such derivatives has also received much attention and in general, many such complexes act as powerful oxidants for a range of organic and

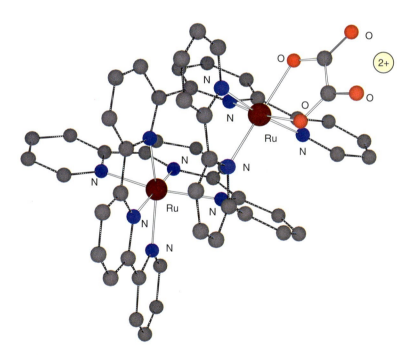

Figure 6.19 *Double-helical structure of $[Ru_2L_2(C_2O_4)]^{2+}$ (L = quinquepyridine,* **35***)*[84]

[83]   C.O. Dietrich-Buchecker, J.-P. Sauvage, N. Armaroli, P. Ceroni and V. Balzani, *New J. Chem.*, 1996, **20**, 801.
[84]   P.K.-K. Ho, K.-K. Cheung and C.-M. Che, *Chem. Commun.*, 1996, 1197.

other species. In a study by the Che group, it proved possible to prepare a 'composite' complex that spans both categories. Thus, oxidation (either chemical or electrochemical) of the double-helical $[Ru_2L_2(H_2O)_2]^{4+}$ (L = quinquepyridine) cation in aqueous solution results in a product containing both a photoactive ruthenium centre and a high-valent Ru=O centre in the one complex.

The achievement of double-helical co-ordination in complexes of the type discussed so far tends to be favoured whenever a conjugated (potentially) polydentate ligand co-ordinates to a metal ion that is too small to allow ready formation of a square planar co-ordination geometry.[81] Nevertheless, this is not necessarily the case when the ligand has additional donor sites available. For example, the moderately large palladium(II) ion has been demonstrated to yield a dinuclear double-helical complex containing two quinquepyridines.[85,86] In this case each metal adopts an irregular five-co-ordinate geometry, showing four short contacts (1.941–2.085 Å) to a 'terpridyl' fragment in one ligand strand and a terminal pyridyl from the other ligand strand. Co-ordination is completed by a long contact (approximately 2.6 Å) to the remaining pyridyl nitrogen of the second ligand. Once again, π-stacking interactions between the two strands are present. While such interactions may aid the stability of the double-helical arrangement in complexes of this type, it is important to emphasise that the relative magnitude of such contributions is difficult to assess and may indeed be rather weak. When quaterpyridine (**36**) was employed, a 1 : 1 complex of type $[PdL][PF_6]_2$ was isolated – presumably incorporating the metal in a square planar geometry.[86]

36

In an early study by Lehn *et al.*, it was also demonstrated that quaterpyridine (**36**) forms a monomeric copper(I) species (that is not helical), whereas with copper(II) a dimeric species is obtained; these complexes may be interconverted electrochemically.[87] The above monomeric copper(I) complex was shown by X-ray diffraction to have a tetragonal pyramidal structure [four basal nitrogens and an axial oxygen (water)] in the solid state. This ligand also yields a dinuclear complex with ruthenium(II) of type $[Ru_2L_2]^{4+}$.

In a related investigation to that just discussed,[88] the tetramethyl-substituted quaterpyridine **37**, which was designed to promote co-ordination in a twisted configuration, yielded complexes with 1 : 1 and 2 : 1 (metal : ligand) stoichiometries with manganese(II) and cobalt(II). With copper(I), a dimeric helical complex of

85   E.C. Constable and M.D. Ward, *J. Am. Chem. Soc.*, 1990, **112**, 1256.
86   E.C. Constable, S.M. Elder, J. Healy, M.D. Ward and D.A. Tocher, *J. Am. Chem. Soc.*, 1990, **112**, 4590.
87   J.-P. Gisselbrecht, M. Gross, J.-M. Lehn, J.-P. Sauvage, R. Ziessel, C. Piccinni-Leopardi, J.M. Arrieta, G. Germain and M. Van Meerssche, *New J. Chem.*, 1984, **8**, 661.
88   J.-M. Lehn, J.-P. Sauvage, J. Simon, R. Ziessel, C. Piccinni-Leopardi, *New J. Chem.*, 1983, **7**, 413.

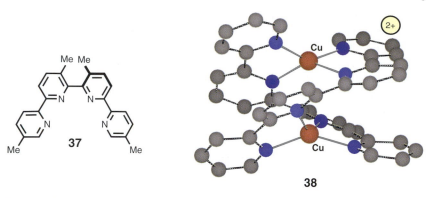

Figure 6.20 *Double-helical structure of [Cu₂L₂]²⁺ (L = 37)*[88]

stoichiometry [Cu₂L₂](ClO₄)₂·H₂O (**38** in Figure 6.20) was isolated. However, Constable *et al.*[89] demonstrated that the parent (unsubstituted) ligand **36** also forms helical dinuclear silver(I) and copper(I) complexes. This latter study served to illustrate that the role of the methyl substituents in **38** was not to control the assembly of the helix but rather to influence its pitch.

The non-symmetrical alkyl-substituted quaterpyridines **39** and **40** (Figure 6.21) also react with an excess of [Cu(MeCN)₄](PF₆) to yield the corresponding complexes of type [Cu₂L₂](PF₆)₂.[90] An analysis of the NMR spectra of each product indicated that, while the ligand derivative with R = Me (**39**) yields a 1 : 1 mixture of 'head-to-head' and 'head-to-tail' dimers (see Figure 6.21), the presence of the

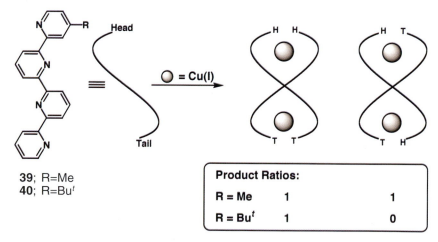

Figure 6.21 *Product ratios for the formation of the head-to-head and head-to-tail isomers of type [Cu₂L₂]²⁺ (**39** or **40**)*[90]

[89] E.C. Constable, M.J. Hannon, A. Martin, P.R. Raithby and D.A. Tocher, *Polyhedron*, 1992, **11**, 2967.

[90] E.C. Constable, F.R. Heirtzler, M. Neuburger and M. Zehnder, *Chem. Commun.*, 1996, 933.

more bulky *tert*-butyl substituent in **40** results in the exclusive formation of the 'head-to-head' isomer. While the origins of the directional specificity in this latter system may involve subtle electronic contributions, it appears that the major influence is the occurrence of steric clashes involving the *tert*-butyl groups if the head-to-tail isomer were to form.

Arising directly from the observation that particular polypyridyl ligands are able to offer a different donor array to each of two metals, Constable *et al.*[91] were successful in constructing a mixed-metal, double-helicate complex of **35**, incorporating both cobalt(II) and silver(I) ions. The synthesis started from helical $[Co_2L_2(OAc)](PF_6)$ (where L = **35**), which had been demonstrated previously to dissolve in donor solvents to yield a mononuclear seven-co-ordinate species of type $[CoL(solvent)_2]^{2+}$ (where L = **35** and solvent = MeCN, MeOH or H_2O). This latter behaviour was rationalised as resulting from the lower stability of the bound cobalt(II) ion originally present at a bipyridyl/bipyridyl site [relative to the second cobalt(II) ion which appears to occupy a terpyridyl/terpyridyl site]. In any case, treatment of the above solution with silver(I) yielded the mixed-metal dinuclear complex ion $[CoAgL_2]^{3+}$. The X-ray structure of this species confirmed that a double-helical structure had been maintained – with the cobalt(II) ion in a 'terpyridyl/terpyridyl' six-co-ordinate environment, while the silver(I) occupies a four-co-ordinate 'bipyridyl/bipyridyl' site (Figure 6.22).

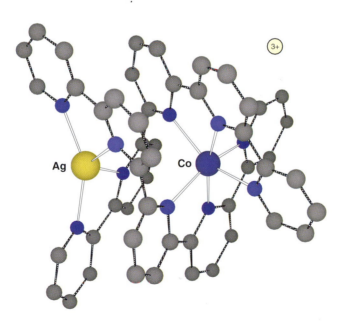

Figure 6.22 *X-Ray structure of the mixed-metal, double-helical cation, [CoAgL₂]²⁺ (L = **35**)*[91]

[91] E.C. Constable, A.J. Edwards, P.R. Raithby and J.V. Walker, *Angew. Chem., Int. Ed. Engl.*, 1993, **32**, 1465.

The six-pyridyl ligand **41**, 'sexipyridine',[85] has been shown to yield complexes with a range of first- and second-row transition metals. Binuclear double-helical cations of type $[M_2L_2]^{4+}$ are formed with manganese(II), iron(II), nickel(II),

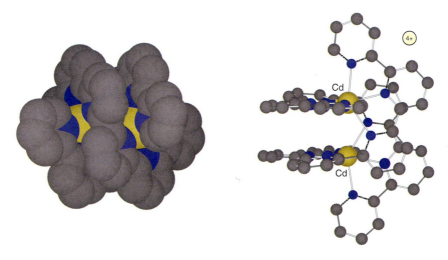

41

copper(II) or cadmium(II); while species of type $[M_3L_2]^{3+}$ occur with copper(I) and silver(I).[92,93]

A binuclear palladium(II) complex with a metal : ligand stoichiometry of 2 : 1 was also obtained.[92]

The preparation of $[Cd_2L_2](PF_6)_4{\cdot}4CH_3CN$ by reaction of cadmium(II) acetate with **41** in methanol, followed by addition of hexafluorophosphate and recrystallisation of the product from acetonitrile, represents a typical synthesis.[92] The X-ray structure of the product confirmed that it has the double-helical geometry shown in Figure 6.23. Each ligand presents a 'terpyridyl' donor arrangement to each cadmium ion. NMR data suggest that the double-helical arrangement found in the solid also persists in solution.

While sexipyridine **41** forms double-helical, binuclear complexes with several transition metal ions (see above), it forms only 1 : 1 complexes with the lan-

Figure 6.23 *The double-helical structure of $[Cd_2L_2]^{4+}$ (L = **41**)*[92]

92   E.C. Constable, M.D. Ward and D.A. Tocher, *J. Chem. Soc., Dalton Trans.*, 1991, 1675.
93   E.C. Constable and R. Chotalia, *J. Chem. Soc., Chem. Commun.*, 1992, 64.

thanides.[94] Thus the photoactive product, [EuL(NO$_3$)$_2$][NO$_3$], is obtained on reaction of this ligand with europium(III) nitrate. The X-ray structure of the product showed that it is a ten-co-ordinate species, with the organic ligand acting as a hexadentate and co-ordinating in a (mono) helically twisted manner while bidentate nitrato ions occupy the remaining co-ordination positions.

In some cases the nature of the substituents present on polypyridine ligands can dramatically affect the geometries adopted on complex formation.[95] For example, 6,6″-dimethyl, and 6,6″-dimethyl together with 4′,4‴-diphenyl substituents on quinquepyridine (**35**) both yield ligands which form double-helical complexes of type [Ag$_2$L$_2$]$^{2+}$. This contrasts with the behaviour of the parent (unsubstituted) ligand which has been reported to form a mononuclear, near-planar (five-co-ordinate) complex with this ion under similar experimental conditions.[96] The methyl substituents in the 'terminal' positions of the substituted ligands apparently sterically inhibit the formation of similar, planar, mononuclear complexes with these systems. In the formation of the helical complexes the substituents influence the assembly process such that, in the first of the above complexes (namely, the derivative with 6,6″-dimethyl substituents), the ligands give rise to the usual [2+3] co-ordination mode – with the silver ions adopting a flattened, distorted trigonal-bipyramidal geometry. However, in the second complex (incorporating both 6,6″-dimethyl and 4′,4″-diphenyl substituents), the ligands yield a [2+1+2] mode in which the central pyridyl nitrogen does not co-ordinate but rather acts as a rigid 'spacer' between the two co-ordination centres. Thus, each silver(I) is four-co-ordinate in this structure. It appears that the increased conjugation in the ligand associated with the presence of phenyl substituents in the 4′,4‴-positions results in a less flexible system which inhibits co-ordination of the central pyridine nitrogen to a silver ion. Both metal complex structures contain π-stacking interactions, with the strength of these being greater in the first complex than in the second.

The introduction of bis(*p*-chlorophenyl) and bis(methylthio) substituents in the 4′- and 4″-positions of quinquepyridine (**35**) (that is, on the backbone of the second and fourth pyridyl rings in the string) was reported to have little effect on the co-ordination behaviour of these ligands relative to that of the unsubstituted analogue (see earlier).[97] Namely, with nickel(II) and copper(I) and/or copper(II), double-helical 2 : 2 complexes were obtained, while cobalt(II) forms both a 1 : 1 and a 2 : 2 complex in the solid state (although, in solution only a mononuclear seven-co-ordinate complex occurs – with the solid state, double-helical structure interconverting to this form upon dissolution).

Terpyridine is the simplest oligopyridine capable of forming a double-stranded helicate, but early attempts to interact this ligand with copper(I) led to air-unstable products that readily oxidised to the well known octahedral bis(terpyridine)cop-

[94]　E.C. Constable, R. Chotalia and D.A. Tocher, *J. Chem. Soc., Chem. Commun.*, 1992, 771.
[95]　Y. Fu, J. Sun, Q. Li, Y. Chen, W. Dai, D. Wang, T.C.W. Mak, W. Tang and H. Hu, *J. Chem. Soc., Dalton Trans.*, 1996, 2309.
[96]　E.C. Constable, M.G.B. Drew, G. Forsyth and M.D. Ward, *J. Chem. Soc., Chem. Commun.*, 1988, 1450.
[97]　E.C. Constable, M.A.M. Daniels, M.G.B. Drew, D.A. Tocher, J.V. Walker and P.D. Wood, *J. Chem. Soc., Dalton Trans.*, 1993, 1947.

per(II) species.[98] Other polypyridines, suitable for use in related studies, have been prepared. For example, Potts *et al.*[99] have reported the synthesis, in moderate to good yields, of an extended range of functionalised polypyridines containing between 2 and 10 linked pyridyl moieties.

Since, as discussed, several oligopyridine-copper(I) double-stranded complexes appear to contain significant π-attraction between adjacent ligand strands, the Potts' group reasoned that the overall π-interaction, and perhaps the stability of the copper(I) complex, would be enhanced by introduction of 6,6″-phenyl substituents onto terpyridine.[100] This was found to be the case. The derivative 6,6″-diphenyl-4,4″-bis(methylthio)-2,2′:6′,2″-terpyridine (**42**) (Figure 6.24) yields an orange-red helical cation of type [Cu$_2$L$_2$]$^{2+}$ whose X-ray structure **43** confirmed that π-interactions involving all aromatic rings in the structure occur; four sets of π-stacks are present. Two sets of triple stacks result from the middle pyridine rings being 'sandwiched' between the two terminal phenyl rings of the second strand, with the average inter-ring distance being 3.80 Å. Two additional sets of double stacks (involving the other two pyridine rings of each strand) are also present.

The bis(methylthio)-substituted quarterpyridine, 4′,4″-bis(methylthio)-2,2′:6′,2″:6″,2‴-quaterpyridine (**44**), yields a binuclear copper(I) complex, but a mononuclear species with copper(II).[101] A spectroelectrochemical study indicated that a redox-induced transformation occurs between these two complex species. The X-ray structure of the 2 : 2 [copper(I) : ligand] helical complex confirmed that each copper is tetrahedrally co-ordinated to four pyridyl nitrogens from two ligand strands, with a short Cu–Cu distance of 3.32 Å being present.

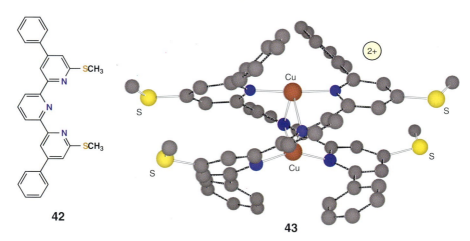

Figure 6.24 *X-Ray structure of the cation [Cu$_2$L$_2$]$^{2+}$ (L = **42**); there are four pairs of π-stacks in this structure*[100]

98   R. Hogg and R.G. Wilkins, *J. Chem. Soc.*, 1962, 341; E.C. Constable, *Adv. Inorg. Radiochem.*, 1986, **30**, 69; J.-V. Folgado, W. Henke, R. Allmann, H. Stratemeier, D. Beltran-Porter, T. Rojo and D. Reinen, *Inorg. Chem.*, 1990, **29**, 2035.
99   K.T. Potts, K.A.G. Raiford and M. Keshavarz-K, *J. Am. Chem. Soc.*, 1993, **115**, 2793.
100 K.T. Potts, M. Keshavarz-K, F.S. Tham, H.D. Abruna and C. Arana, *Inorg. Chem.*, 1993, **32**, 4450.
101 K.T. Potts, M. Keshavarz-K, F.S. Tham, H.D. Abruna and C.R. Arana, *Inorg. Chem.*, 1993, **32**, 4422.

44

In this same study, alkylthio-substituted quinquepyridines were demonstrated to form single- and mixed-valent, binuclear and trinuclear helical complexes in quantitative yield.[101] The presence of the alkylthio substituents was found to aid ligand solubility. The trinuclear copper(I) complex of the above ligand in acetonitrile showed three irreversible, well separated one-electron oxidations. These are attributable to the presence of three copper(I)/(II) couples. The significant difference between the oxidation potentials for these couples was taken as evidence for the presence of a metal–metal interaction in this homonuclear complex. The corresponding double-helical copper(II) species is of type $[Cu_2L_2(OAc)][PF_6]_3$ and contains the individual copper atoms in different co-ordination environments (N_6 and N_4O_2), with a supplementary bidentate acetate ligand completing the co-ordination shell of one of the copper ions.

The disubstituted polypyridine derivatives, 4′,4‴-bis(methylthio)-2,2′:6′,2″:6″, 2‴:6‴,2⁗-quinquepyridine (**45**) and 4′,4⁗-bis(methylthio)- 2,2′:6′,2″:6″,2‴:6‴,2⁗: 6⁗,2⁗′-sexipyridine (**46**), in which the alkylthio groups are present on the 4-positions of the penultimate pyridines (with respect to each end) also form bimetallic double-stranded helical complexes with iron(II), cobalt(II), nickel(II), zinc(II) and palladium(II).[102] With the first of these ligands, the palladium species is of type

45

46

$[M_2L_2]^{4+}$ whereas the other metals yield complexes of type $[M_2L_2(OAc)]^{3+}$. The X-ray structure of $[Ni_2L_2(OAc)]^{3+}$ confirmed its double-helical nature, with the metal ions showing distorted octahedral geometries and, once again, different co-ordination environments (N_6 and N_4O_2).

Further studies, including electrochemical redox-state transformations, involving other double-stranded helicates derived from analogously functionalised quater-pyridine, quinquepyridine and septipyridine, have also been reported.[101,103]

[102] K.T. Potts, M. Keshavarz-K, F.S. Tham, H.D. Abruna and C. Arana, *Inorg. Chem.*, 1993, **32**, 4436.
[103] K.T. Potts, M. Keshavarz-K, F.S. Tham, K.A.G. Raiford, C. Arana and H.D. Abruna, *Inorg. Chem.*, 1993, **32**, 5477.

As discussed already, when quinquepyridine **35** forms a double-stranded, dinuclear helicate it most commonly presents bipyridyl and terpyridyl domains to the respective bound metal ions. Constable *et al.*[104] reasoned that replacement of the central pyridine of quinquepyridine by a 1,3-phenylene group would result in a similarly rigid ligand which incorporated spaced bipyridyl domains for binding to two metal centres. In confirmation of this, the rigidly-spaced bis-bipyridyl system **47** (Figure 6.25) reacts with nickel acetate in methanol to yield a crystalline blue-green helicate of type $[Ni_2(OAc)_2L_2][PF_6]_2$ upon addition of hexafluorophosphate anion.[104,105] As anticipated, each nickel centre is co-ordinated to two 2,2-bipyridyl fragments (one from each ligand). A bidentate acetate group completes each metal's co-ordination sphere. The respective bipyridyl moieties are slightly twisted, but the major twisting occurs between these groups and the 1,3-phenylene spacer. Interestingly, no face-to-face stacking interactions are present between the near co-planar aromatic rings in this system.

The polypyridine derivatives **48** and **49** also yield double-helical complexes of type $[Cu_2L_2]^{4+}$ (L = **48** or **49**) and $[Ag_2L_2]^{2+}$ (L = **48** or **49**) as well as the single-helical species, $[Ru_2L(terpy)_2]^{4+}$ (where L = **49** and terpy is 2,2':6',2''-terpyridine).[106] The electrochemistry of the double-helical cations indicates substantial electronic coupling between the metals in each case.

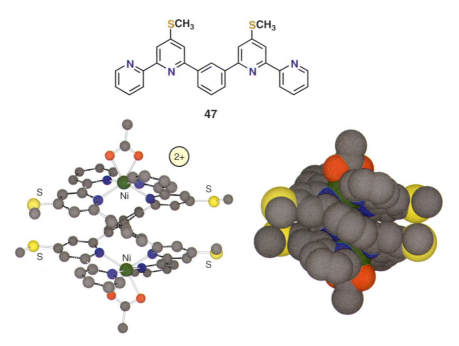

47

Figure 6.25 *Structure of the double-helical cation $[Ni_2(OAc)_2L_2]^{2+}$ (L = **47**)*[104]

104 E.C. Constable, M.J. Hannon and D.A. Tocher, *J. Chem. Soc, Dalton Trans.*, 1993, 1883.
105 E.C. Constable, M.J. Hannon and D.A. Tocher, *Angew. Chem., Int. Ed. Engl.*, 1992, **31**, 230.
106 P.K.-K. Ho, S.-M. Peng, K.-Y. Wong and C.-M. Che, *J. Chem. Soc., Dalton Trans.*, 1996, 1829.

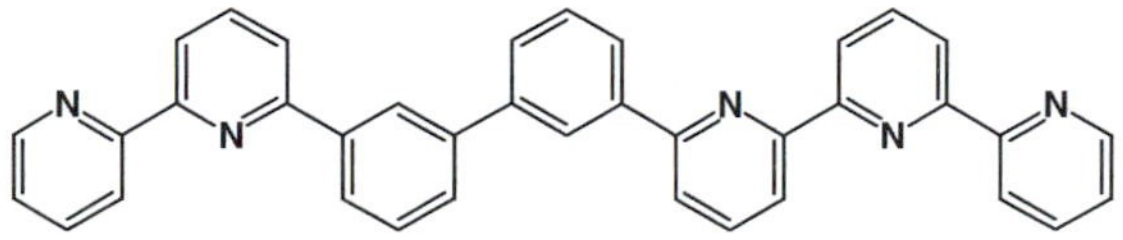

48

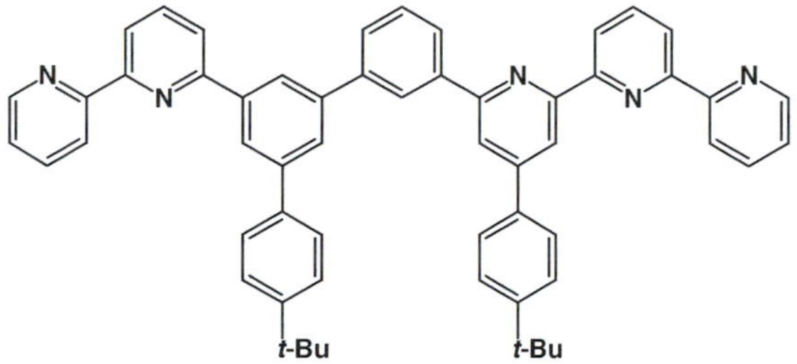

t-Bu   *t*-Bu

49

The redox-active derivative, 4′,4″″-diferrocene-functionalised sexipyridine **50** also yields double helical complexes of type $[M_2L_2]^{4+}$ (M = Fe, Co, Ni or Zn).[107] The presence of the ferrocene substituents does not appreciably affect the co-ordination behaviour towards these metal ions. As expected, all of these complexes are electrochemically active, giving rise to a variety of redox behaviour (which in some cases has not been able to be unambiguously assigned to specific processes). However, in general, it was observed that the reversibility of redox processes associated with the metal centres co-ordinated to individual 'terpyridyl' domains is lost if these processes follow ferrocene-centred Fe^{II}–Fe^{III} redox reactions. Related studies involving 4′,4″′-bis(ferrocenyl)-2,2′:6′,2″:6″,2″′:6″′,2″″-quinquepyridine,

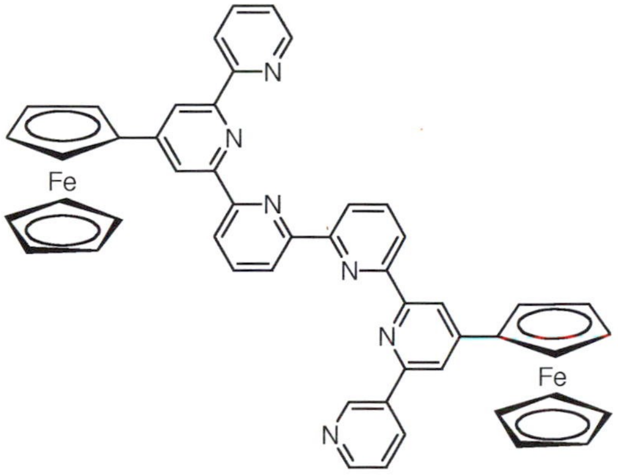

50

[107] E.C. Constable, A.J. Edwards, R. Martinez-Mánez and P.R. Raithby, *J. Chem. Soc., Dalton Trans.,* 1995, 3253.

which forms double-helical complexes with both nickel and copper, have also been reported.[108]

6.5.3 Systems Incorporating Segmented Terpyridyl Derivatives

A segmented tris-terpyridyl ligand **51** (Figure 6.26), incorporating three terpyridyl domains linked by $-(CH_2)_2-$ bridges has been employed for the preparation of double-helical complexes. The reaction of iron(II) or nickel(II) with this ligand leads to self-assembly of the corresponding $[M_3L_2]^{6+}$ complexes of type **52** and **53**[109] (Figure 6.26) – the iron(II) helicate has been resolved into its enantiomers.

Based on the well documented observation that metal ions favouring tetra-co-ordination [such as copper(I) or silver(I)] or hexa-co-ordination [such as iron(II) or nickel(II)] preferentially bind to two bipyridyl or two terpyridyl units, respectively. Lehn *et al.*[110] considered that one bipyridyl unit and one terpyridyl unit arranged more or less mutually perpendicular might favour co-ordination to copper(II) – a metal ion which is frequently penta-co-ordinated. Hence, it seemed feasible that an oligo-bipyridyl strand and an oligo-terpyridyl strand might give rise

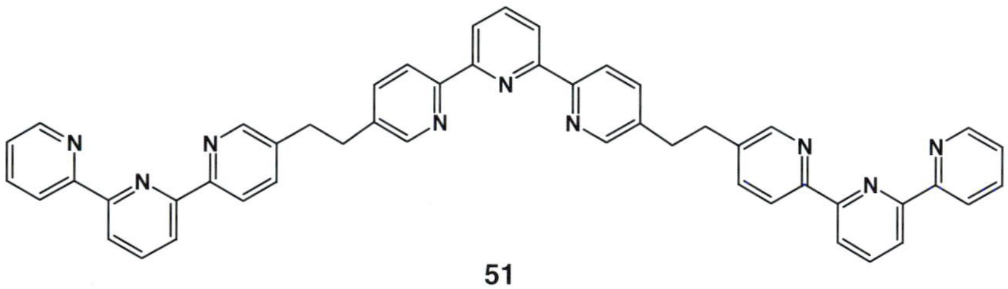

51

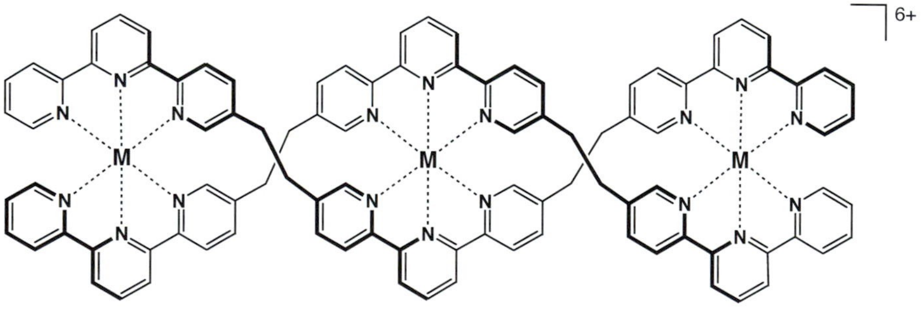

52; M=Fe(II)
53; M=Ni(II)

Figure 6.26 *Double-helical structures of the complexes* **52** *and* **53** *of type* $[M_3L_2]^{6+}$ *[M = Fe(II), Ni(II)] where L is the 'segmented' ligand* **51**[109]

108 E.C. Constable, R. Martinez-Mánez, A.M.W.C. Thompson and J.V. Walker, *J. Chem. Soc., Dalton Trans.*, 1994, 1585.
109 B. Hasenknopf and J.-M. Lehn, *Helv. Chim. Acta*, 1996, **79**, 1643.
110 B. Hasenknopf, J.-M. Lehn, G. Baum and D. Fenske, *Proc. Natl. Acad. Sci. USA*, 1996, **93**, 1397.

to a mixed-strand double helicate in which co-ordination to five-co-ordinated copper(II) occurs at each metal binding site. In this regard, it is noted that the monomeric species [Cu(bipyridine)(terpyridine)]$^{2+}$ contains the copper in a square pyramidal environment;[111] however, it is also noted that a trigonal bipyridyl arrangement is also common for copper(II).

Reaction of copper(II) with a 1 : 1 mixture of the ligand incorporating three link-bipyridine groups (**54**) and the analogue mentioned above, incorporating three linked 2,2′:6′,2″-terpyridine groups, yielded a linear trinuclear complex in which the two different ligand strands wrap around three penta-co-ordinated copper(II) ions in a helical fashion.[110] The co-ordination geometry of the central copper(II) in this product is trigonal bipyramidal while the two outside copper(II) ions are in similar square pyramidal environments.

54

square pyramidal   trigonal bipyramidal   square pyramidal

6.5.4   Systems Based on Benzimidazole and Related Ligands

A range of studies based on benzimidazole-containing or related polyfunctional ligands have led to the generation of a considerable number of new helical complexes. In some ways these investigations parallel the polypyridyl- and poly-bipyridyl-based studies discussed already. An early study involved the bis(methyl-benzimidazole) derivative **55**, which was demonstrated to form a double-helical complex of type [Cu$_2$L$_2$]$^{2+}$.[112] An unusual feature of this structure is that the pair of ligands act essentially as a bis(monodentate) system towards each copper ion, giving rise to quasi-linear co-ordination. However, each metal also has a weak interaction with a 'bridging' pyridine such that its co-ordination geometry is perhaps

[111]   E. Baum, (1987) Diplomarbeit thesis (Philipps Universitat, Marburg, Germany); G. Arena, R.P. Bonomo, S. Musumeci, R. Purrello, E. Rizzarelli and S. Sammartano, *J. Chem. Soc., Dalton Trans.,* 1983, 1279.
[112]   C. Piguet, G. Bernardinelli and A.F. Williams, *Inorg. Chem.*, 1989, **28**, 2920.

best considered to lie somewhere between linear and a highly distorted tetrahedral. Intramolecular stacking between the aromatic planes of the heterocyclic ligands is clearly evident in this structure.

55   **56**

In order to investigate the importance of intramolecular stacking, the analogous imidazole ligand **56**, containing two less aromatic rings, was substituted for **55** in the preparation of the corresponding metal complex.[113] Although the X-ray crystal structure of the resulting product confirmed that the basic $[Cu_2L_2]^{2+}$ unit is pre-served, intramolecular stacking was no longer present. One ligand in this complex is rotated by approximately 90° about the two-fold axis with respect to the other (when compared with the corresponding complex of **55**). Interestingly, this permits strong intermolecular stacking between the imidazole planes of *different* complex-es (average interplanar distances of 3.48 and 3.45 Å) so that the crystal structure approximates an infinite double helix. Thus, the loss of *intramolecular* stacking has been compensated by formation of *intermolecular* stacking between individual com-plex units in this system. Complete separation of complexes with different helici-ty occurs in the crystal state, with the mirror image forms stacking alongside one another in parallel columns.

In a further study, the related ligand 1,3-bis(1-methylbenzimidazol-2-yl)benzene, in which the central pyridine of **55** is replaced by a benzene moiety, has been used to form the corresponding dinuclear copper(I) species.[114] In this complex, each cop-per is linearly co-ordinated by a benzimidazole-nitrogen originating from each lig-and. In contrast to the related complex of **55**, the absence of a central pyridine donor in this complex is sufficient for it to revert to a non-helical geometry in which a 'side-by-side', non-intertwined arrangement of its two ligands is present. In this structure there is evidence for the presence of a weak stacking interaction between the parallel bridging phenyl groups linking the two metal binding sites. It is noted that synthetic procedures for obtaining higher-order, mixed pyridyl–benzimidazole ligands for use in the synthesis of new helical species have also been developed.[115]

The development of new segmented ligands with non-identical binding sites, such as **57–59**,[116,117] has led to the formation of a number of both mononuclear and din-uclear, helicates that include mixed-metal species. For example, interaction of **58**

[113] R.F. Carina, G. Bernardinelli and A.F. Williams, *Angew. Chem., Int. Ed. Engl.*, 1993, **32**, 1463 and references therein.

[114] S. Ruttimann, C. Piguet, G. Bernardinelli, B. Bocquet and A.F. Williams, *J. Am. Chem. Soc.*, 1992, **114**, 4230.

[115] C. Piguet, B. Bocquet and G. Hopfgartner, *Helv. Chim. Acta*, 1994, **77**, 931.

[116] A.F. Williams, C. Piguet and G. Bernardinelli, *Angew. Chem., Int. Ed. Engl.*, 1991, **30**, 1490.

[117] C. Piguet, G. Hopfgartner, A.F. Williams and J.-C.G. Bünzli, *J. Chem. Soc., Chem. Commun.*, 1995, 491; C. Piguet, G. Hopfgartner, B. Bocquet, O. Schaad and A.F. Williams, *J. Am. Chem. Soc.*, 1994, **116**, 9092.

with iron(II), cobalt(II) or zinc(II) in acetonitrile yields mononuclear 'head-to-head' complexes of type $[ML_2]^{2+}$. In these, two tridentate domains of **58** co-ordinate to the respective metals to yield an octahedral co-ordination arrangement in each case. It has been demonstrated by ^1H NMR studies that such complexes are able to inter-act further with a second metal ion in acetonitrile to yield homo-dinuclear, double-helical complexes of type $[M_2L_2]^{4+}$. In these, the second cation is bound in the pseudo-tetrahedral site.

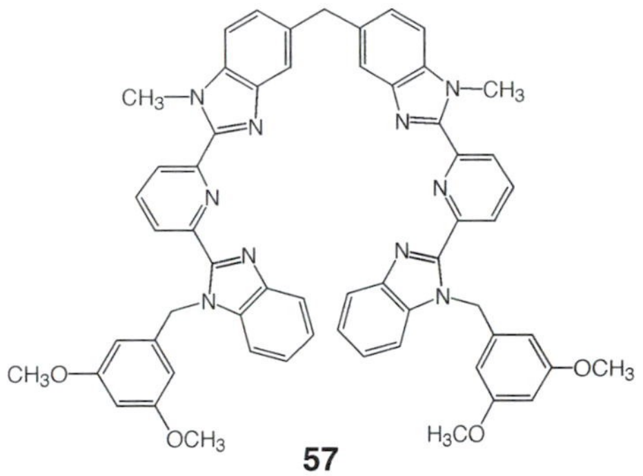

CH₃O—

57

H₃CO

CH₃O—

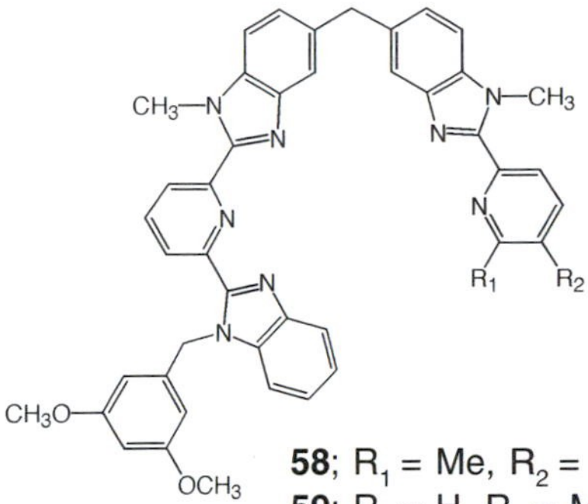

58; R$_1$ = Me, R$_2$ = H
59; R$_1$ = H, R$_2$ = Me

The above mono-nuclear iron(II) complex reacts with silver(I) to yield the hetero-dinuclear species, $[FeAgL_2]^{3+}$. In this, the iron(II) is bound in an octahedral site while the silver(I) ion occupies a tetrahedral one. The success of this synthe-sis thus depends upon a ligand design that allows 'programmed' co-ordination of the respective metal ions such that each attains its preferred stereochemistry.

Baker *et al.*[118] have demonstrated that the 2,6-pyridyl-bridged species **60** reacts with copper(I) to yield a complex of type $[Cu_3L_2(CH_3CN)_2](PF_6)_3$. The X-ray struc-

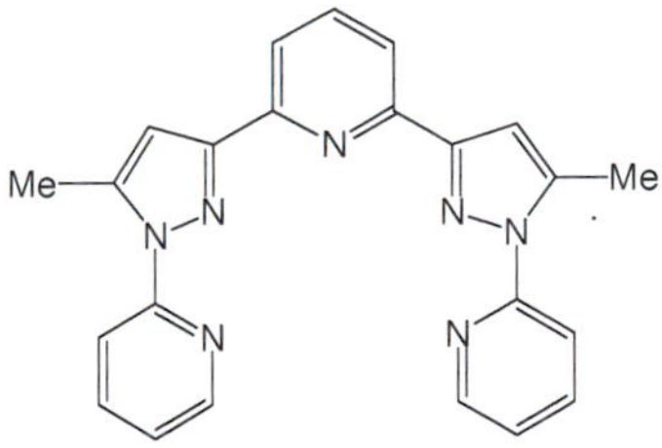

60

ture of this product confirms the double-helical geometry of the cation and reveals that two unique copper(I) sites are present in a 2 : 1 ratio. All three copper atoms achieve tetrahedral co-ordination with the central copper being co-ordinated by two pyridylpyrazole fragments (corresponding to the third and fourth heterocycles of each ligand string), while the remaining copper(I) ions are co-ordinated by a pyridylpyrazole fragment (composed of the first and second heterocycles of the

118 A.T. Baker, D.C. Craig and G. Dong, *Inorg. Chem.*, 1996, **35**, 1092.

ligand string) as well as a pyridyl group (the fifth heterocycle from the second ligand string) and an acetonitrile ligand.

6.5.5 Helicates Derived from Macrocyclic Ligand Systems

A number of semi-flexible, large-ring macrocyclic ligands have been demonstrated to yield helical metal complexes.[119–123] For example, the Schiff base derivative **61** was postulated to adopt helical co-ordination geometries with a number of metal ions.[121] The X-ray structure of [PbL]$^{2+}$ (L = **61**) is shown in Figure 6.27. The double-helical array is stabilised in this case by no less than five aromatic π-interactions. NMR evidence suggests that the helical arrangement persists in solution.

X-Ray structures of the corresponding cobalt(II), nickel(II) and zinc(II) deriva-tives of **61** confirm that each of these complexes adopts a related helical geometry in which the co-ordination sphere is filled by six nitrogen donors from two essen-tially planar (meridionally co-ordinated) pyridine-2,6-diyldiimine units belonging to the macrocycle.[124]

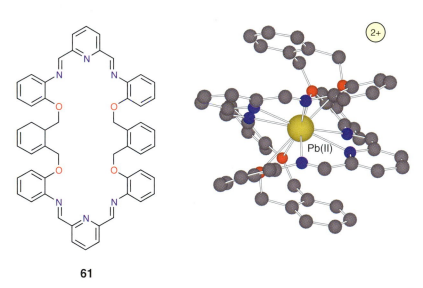

61

Figure 6.27 *Structure of the double-helical, macrocyclic complex, [PbL]$^{2+}$ (L = **61**)[121]*

[119] S.W.A. Bligh, N. Choi, E.G. Evagorou, W.-S. Li and M. McPartlin, *J. Chem. Soc., Chem. Commun.*, 1994, 2399.
[120] P. Comba, A. Fath, T.W. Hambley and D.T. Richens, *Angew. Chem., Int. Ed. Engl.*, 1995, **34**, 1883.
[121] R.W. Matthews, M. McPartlin and I.J. Scowen, *Chem. Commun.*, 1996, 309; D.E. Fenton, R.W. Matthews, M. McPartlin, B.P. Murphy, I.J. Scowen and P.A. Tasker, *J. Chem. Soc., Dalton Trans.*, 1996, 3421.
[122] P.M. Fitzsimmons and S.C. Jackels, *Inorg. Chim. Acta*, 1996, **246**, 301.
[123] P. Comba, A. Fath, T.W. Hambley and A. Vielfort, *J. Chem. Soc., Dalton Trans.*, 1997, 1691.
[124] D.E. Fenton, R.W. Matthews, M. McPartlin, B.P. Murphy, I.J. Scowen and P.A. Tasker, *J. Chem. Soc., Chem. Commun.*, 1994, 1391.

Ligands **62** and **63** have been demonstrated to yield related helical complexes of type $[Cu_2L]^{2+}$.[120] Both complexes correspond to the 'figure of eight' motif illustrated by **64**. The adoption of this arrangement appears to reflect the preferred tetrahedral co-ordination of the respective copper(I) ions coupled with the ability for

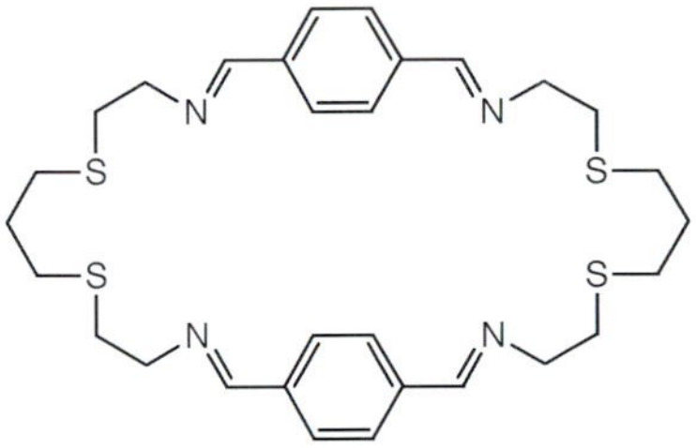

62 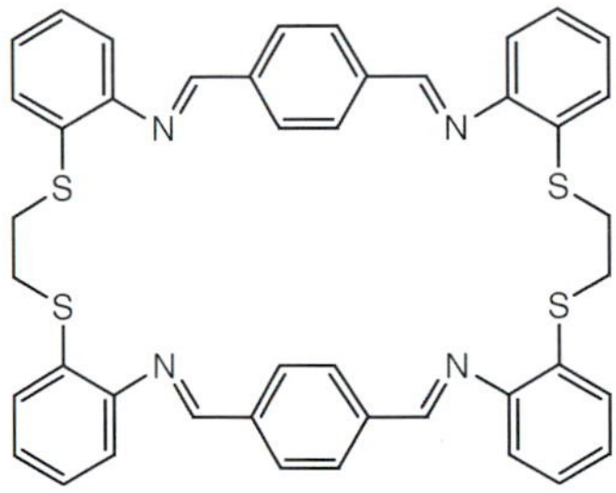   **63**

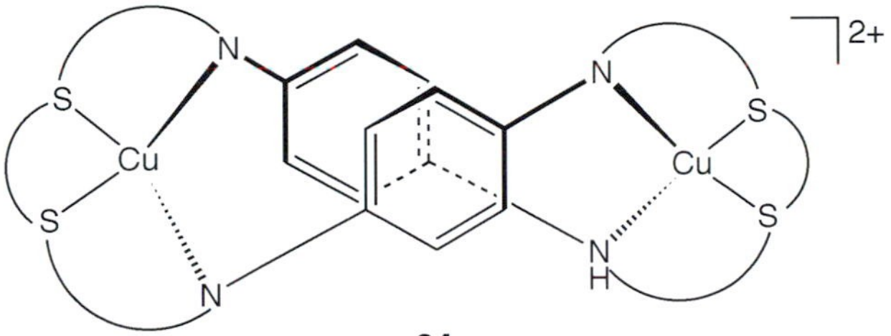

64

π-stacking of the linking phenylene rings to take place. These rings adopt a parallel arrangement with an 'ideal' separation of approximately 3.5 Å in each complex.

6.5.6 Systems Based on Non-transition Metal Templates

Investigations of the present type have not been restricted to systems incorporating only transition and post-transition ions. For example, 'preorganised' helical ligands have been used for the production of alkali metal double-helical structures. Bell and Jousselin[125] employed the 'heterohelicenes' **65** and **66** to yield double-helical 2 : 2 complexes of sodium. These molecular 'coils' are less flexible than the oligopyridines and exist in solution as equal mixtures of right- and left-handed helices. From physical measurements and inspection of CPK models, the structure of the sodium complex, $[Na_2L_2]^{2+}$ (L = **65**), was proposed to incorporate two intertwined molecular coils, with the sodium ions being bound inside the resulting double helix. The tight binding of the metal ions in this arrangement is undoubtedly reflected by the observed substantial stability of this dinuclear complex.

---

[125] T.W. Bell and H. Jousselin, *Nature*, 1994, **367**, 441.

65   **66**

Ligand **67** (Figure 6.28), incorporating two potentially tridentate chelating seg-ments linked by an anionic $-BH_2-$ bridge, reacts with potassium to yield a neutral, double-helical species of type $[K_2L_2]$.[126] An X-ray structure of this product, crys-tallised from chloroform, confirmed its double-helical geometry. Each potassium ion is bound in an irregular six-co-ordinate environment by two tridentate segments from each ligand strand in the usual manner. Once again, the structure contains π-stacking interactions between near-parallel, overlapping aromatic sections of the two ligands.

In a further study involving alkali metals, Cram and co-workers[127] reported the metal binding properties of the two helically chiral ligands, **68** and **69**. These ligand

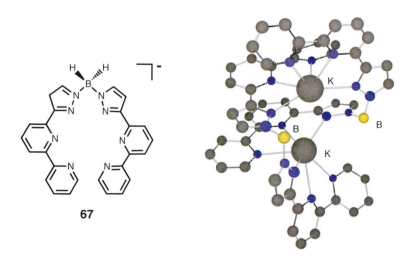

67

Figure 6.28 *The double-helical structure of* $[K_2L_2]$ *(L = **67**)*[126]

[126] E. Psillakis, J.C. Jeffery, J.A. McCleverty and M.D. Ward, *Chem. Commun.*, 1997, 479.
[127] J.K. Judice, S.J. Keipert and D.J. Cram, *J. Chem. Soc., Chem. Commun.*, 1993, 1323.

68   **69**

systems incorporate two phenanthroline units joined to one another through one and two 1,1'-binaphthyl units, respectively. In the case of **69**, the two phenanthroline domains provide four donor nitrogens that are approximately tetrahedrally arranged – a structure imposed by the presence of the chiral binaphthyl groups that link the phenanthrolines. For **68**, a similar arrangement, while not imposed, can readily be adopted. Both ligands form complexes with lithium, sodium and potassium; however, there appears to be little discrimination between the binding of these ions. As expected, the increased preorganisation of **69** over **68** results in stronger metal ion complexation in the former case.

6.5.7   Controlling the Helicity

In a typical study of the type discussed so far, the use of achiral ligands and metal cations results in helical products that are chiral but are formed as racemic mixtures. The question of using (resolved) asymmetric ligand systems for the diastereoselective formation of helicates has been addressed by a number of groups.

In one such study, asymmetric induction was employed to generate complexes of predominantly single-handed helicity.[128] The approach employed was to introduce chiral centres into the backbone of each of the ligand strands. The tris-bipyridine ligand **70**, incorporating two asymmetric carbon centres (with *S,S*-configurations), was employed. On complexation of this species with silver(I) or copper(I), helical products of type $(M_3L_2)^{3+}$ were generated. An investigation of the copper(I) species indicated that there is preferential generation of the right-handed double helicate in this case.[128]

70

[128]   W. Zarges, J. Hall and J.-M. Lehn, *Helv. Chim. Acta*, 1991, **74**, 1843.

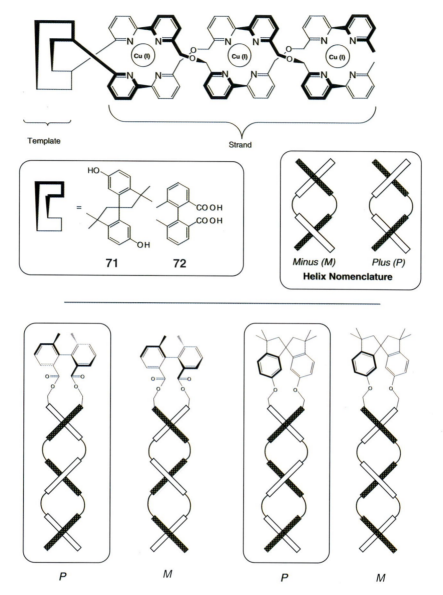

Figure 6.29 *Use of chiral template to direct the helicity of a double-stranded copper(I) helicate*[129]

Lehn *et al.*[72] have described the helical self-assembly process in the following terms:

In the general context of programmed supramolecular systems, helicate formation may be described as the result of the reading by metal ions of the molecular information stored in the oligo(bipyridine) strands following a tetrahedral co-ordination algorithm.

Namely, the structural information 'stored' in each ligand strand is 'read out' upon interaction with each respective metal ion. In the above case, the result is the formation of a supramolecular assembly of defined helicity.

In another study,[129] a chiral templating procedure was employed to direct the helicity of an oligo-bipyridine, double-stranded copper(I) helicate. The strategy employed is illustrated in Figure 6.29 in which spiro-bisindanol **71** and dimethyl-biphenic acid **72** were employed as the chiral template 'starters' for the twisting process inherent in helicate formation. The above approach serves as a model for a general route towards the synthesis of enantio-pure helicates and is also of considerable intrinsic interest since it illustrates the manner by which stereochemical information may be transmitted over nanometer distances.

Subsequently, chiral terpyridines bearing bornyl substituents in their 6-positions (**73** and **74**) have been employed for copper(I) helicate formation.[130] The two enantiomeric forms of the ligand preferentially form dinuclear double helicates with *P* and *M* chiralities, respectively. Both chiral isomers have been characterised structurally by X-ray diffraction.

73   **74**

In another approach, it was demonstrated by X-ray diffraction that the optically pure, tertiary phosphine ligand $(S,S)-(+)-Ph_2PCH_2CH_2P(Ph)CH_2CH_2P(Ph)CH_2CH_2-PPh_2$ yields a left-handed double-helical complex of di-silver(I) as well as a corresponding 'side-by-side' conformer.[131] A molecule of each species is found (together with their associated hexafluorophosphate anions) in each unit cell.

The interaction of the chiral quaterpyridine **75**, bearing fused chiral groups in the 5,6- and 5''',6'''-positions, with copper(I) has been investigated. The respective ligand isomers were demonstrated to form structurally characterised *P* or *M* dinuclear double helicates in approximately 99% diasteromeric excess upon coordination to the above ion.[132]

[129]   C.R. Woods, M. Benaglia, F. Cozzi and J.S. Siegel, *Angew. Chem., Int. Ed. Engl.*, 1996, **35**, 1830.
[130]   E.C. Constable, T. Kulke, M. Neuburger and M. Zehnder, *Chem. Commun.*, 1997, 489.
[131]   A.L. Airey, G.F. Swiegers, A.C. Willis and S.B. Wild, *J. Chem. Soc., Chem. Commun.*, 1995, 695.
[132]   E.C. Constable, T. Kulke, G. Baum and D. Fenske, *Inorg. Chem. Commun.*, 1998, **1**, 80.

75

6.5.8 A Helical System Leading to a Multi-intertwined [2]-Catenane

A metal-ion template (Figure 6.30) approach has been employed for the synthesis of a '4-crossing' [2]-catenane **81** as well as its topological isomer, the corresponding singly-interlocked [2]-catenane (**82**).[133–135] The trinuclear complex **76** was obtained by self-assembly of the organic components **78** and **79** in the presence of copper(I). An attempted ring-closing reaction to yield **77** (under conditions of moderate dilution) using diiodo-heptoethyleneglycol in *N,N*-dimethylformamide in the presence of caesium carbonate, resulted instead in a mixture of copper-containing products. From these, only the mononuclear copper(I) complex **80** was isolated pure (in 13% yield). Because of the difficulty in separating the copper-containing species, the above mixture was treated with cyanide ion and the resulting demetallated products subjected to further chromographic separation. By these means, **78** was recovered in 17% yield together with two topological isomers, **81** (the required product) in 2% yield and **82** in 1% yield.

Finally, it needs to be noted that while a simple [2]-catenane can be made topologically chiral by orientation of the rings – see **83**, a four-crossing [2]-catenane such as **77** is automatically chiral without the need for such orientation.

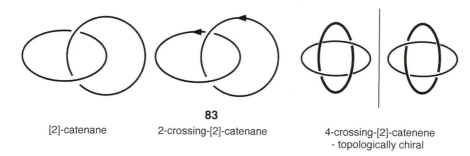

83

[2]-catenane   2-crossing-[2]-catenane   4-crossing-[2]-catenene
- topologically chiral

133   J.-F. Nierengarten, C.O. Dietrich-Buchecker and J.-P. Sauvage, *J. Am. Chem. Soc.*, 1994, **116**, 375.
134   C. Dietrich-Buchecker, E. Leize, J.-F. Nierengarten, J.-P. Sauvage and A.V. Dorsselaer, *J. Chem. Soc., Chem. Commun.*, 1994, 2257.
135   J.-F. Nierengarten, C.O. Dietrich-Buchecker and J.-P. Sauvage, *New J. Chem.*, 1996, **20**, 685.

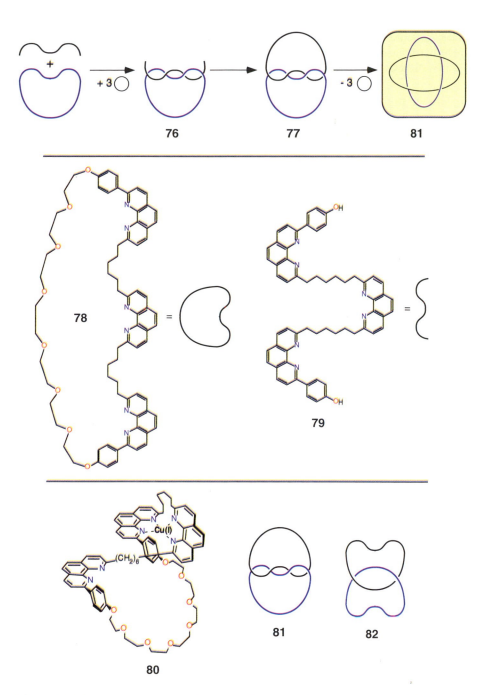

Figure 6.30 *Stepwise strategy for obtaining a 4-crossing-[2]-catenane of type* **81** (top). *The phenanthroline-containing fragments employed* (centre) *and three products from the reaction* (bottom)[133–135]

6.6 Triple Helicates

The stereochemical information stored in an octahedral metal centre has been tapped in order to generate triple helicates. The principle employed is a direct extension of that discussed for the production of double helicates. Most commonly, it involves the successive binding of octahedral metal centres to three bidentate ligand sites belonging to three different di- or poly-bidentate ligand strands.[136]

6.6.1 Transition and Post-transition Metal Systems

In early studies, groups led by Shanzer,[137] Williams[116] and Lehn[66] successfully used octahedral metal ions for the production of triple-helical species.

Shanzer *et al.*[137] reported the preparation of a di-iron(III) triple helix based on the metal-directed intermolecular twisting of the tripodal ligand **84**. This product is stabilised by inter-strand hydrogen bonds which also influence the generation of the diastereomeric helices with respect to their left- or right-handedness. The ligand was derived from a tris(2-aminoethyl)amine bridgehead, with each strand being

84

extended by attachment of L-leucine and further elongated by an alternating sequence of hydroxamate and amide groups. The latter respectively act as ligands for iron(III) and as the (inter-strand) hydrogen bonding units. Thus, as mentioned above, a feature of this system (as well as of other related helical iron-containing derivatives)[138] is the presence of weak, non-covalent interactions which act to stabilise the overall helical topology. Furthermore, it is the presence of asymmetric carbon atoms in the ligand strands that serve to direct the chiral sense of the helix generated.

In a now classic study, Williams *et al.*[116] employed the relatively rigid, (potentially) bis-bidentate ligand **85** (Figure 6.31), incorporating linked benzimidazolyl units, to produce a triple helicate of stoichiometry $[Co_2L_3]^{4+}$. The X-ray structure of this orange-red product is also shown in the figure. Each cobalt(II) ion has a slightly distorted octahedral geometry and is separated from its partner by 8.43 Å. In order to evaluate the effect of the 'terminal' di-methyl substituents on the nature of the complex formed,[139] the structure of the above complex was compared with that of the closely-related ligand **86**. Like **85**, **86** also yields a dinuclear triple-helical cation of type $[Co_2L_3]^{4+}$ in which the bidentate domains are wrapped around

[136]   E.C. Constable, *Angew. Chem., Int. Ed. Engl.*, 1991, **30**, 1450.
[137]   J. Libman, Y. Tor and A. Shanzer, *J. Am. Chem. Soc.*, 1987, **109**, 5880.
[138]   A.M. Albrecht-Gary, J. Libman and A. Shanzer, *Pure Appl. Chem.*, 1996, **68**, 1243.
[139]   C. Piguet, G. Bernardinelli, B. Bocquet, O. Schaad and A.F. Williams, *Inorg. Chem.*, 1994, **33**, 4112.

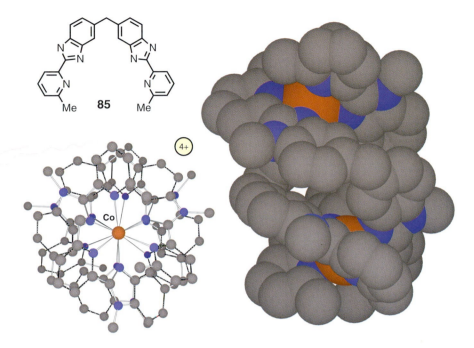

Figure 6.31 *The triple-helical structure of type [Co$_2$L$_3$]$^{4+}$ (L = **85**)[116]*

86

the two cobalt ions which lie on a pseudo-C_3 axis. The co-ordination geometry of each cobalt ion is best described as octahedral, flattened along the C_3 axis. It is clear that steric clashes between the methyl groups occupying the 6-positions of the pyridyl rings of **86** are reflected in increased distortion of the co-ordination spheres around cobalt(II), with a concomitant increase in the respective Co–N(pyridine) bond lengths being evident.

The failure to achieve electrochemical oxidation of this latter cobalt(II) complex [to its cobalt(III) state] has been ascribed to the lower overall stability of this species. This result raises the prospect of employing such subtle steric effects in ligand design for the rational tuning of redox behaviour.

A comparison of the lability of the dinuclear complex [Co$_2$L$_3$]$^{4+}$ (L = **86**) with that of the analogous mononuclear complex of type [CoL$_3$]$^{2+}$ (where L is the 'monomeric' bidentate ligand **87**) has been reported.[140] As might be expected, the 'stitching' together of the bis-bidentate ligand strands by two metal ions leads to a marked decrease in complex lability. While the complex of **87** undergoes rapid isomerisation between meridional and facial forms in acetonitrile, the process for the dinuclear complex is approximately 10^5 times slower. The inertness in the latter system may be attributed to the inherent rigidity of the ligand strands in this complex, coupled with the tight pitch of its helix.

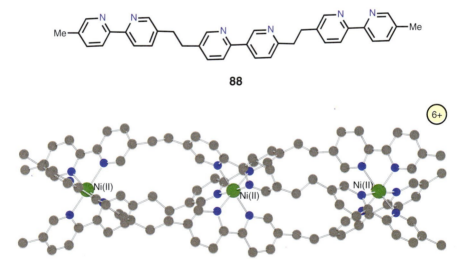

87

Following an earlier study in which a triply-branched dinuclear iron(II) complex of a bis(bipyridyl) ligand was observed to show triple helical features,[66] Lehn and co-workers[141] investigated the use of oligo(bipyridyl) strands for promoting triple helicity in the presence of an octahedral metal ion. Accordingly, this group demonstrated the spontaneous self-assembly of a species of type $[Ni_3L_3]^{6+}$ using the di-substituted, tris(bipyridyl) ligand strand **88**. The X-ray structure of the product is shown in Figure 6.32.

Ligand **89**, consisting of two bidentate bipyridyl moieties linked by a diamide-containing spacer, has been demonstrated to interact with iron(II) to form a water-

88

Figure 6.32 *Structure of the trinuclear complex $[Ni_3L_3]^{6+}$ (L = **88**)*[141]

[140] L.J. Carbineer, A.F. Williams, U. Frey, A.E. Merbach, P. Kamalaprija and O. Schaad, *J. Am. Chem. Soc.*, 1997, **119**, 2488.
[141] R. Krämer, J.-M. Lehn, A. De Cian and J. Fischer, *Angew. Chem., Int. Ed. Engl.*, 1993, **32**, 703.

soluble, tris-helical complex of stoichiometry $[Fe_2L_3]^{4+}$.[142] The ^1H and ^{13}C NMR spectra of this racemic product are in accord with the presence of global D_3 symmetry – as expected for a helical, triply-bridged binuclear complex.

89

The X-ray structure of another helical complex incorporating iron(II) [and iron(III)] is given in Figure 6.33.[143] This unusual mixed valence species is of type $[Fe_3OL_3]$ and contains the doubly-deprotonated form of **90** such that the overall complex is neutral. The 'core' of the complex is an Fe_3–O^{2-} ion which is linked to (formally) one iron(II) and two iron(III) atoms that are arranged at the corners of an equilateral triangle (with the oxygen atom in its centre). The tetrazolyl donors of the organic ligands bind to neighbouring iron atoms from opposite sides of the Fe_3-plane, thus generating the triple-helical geometry. Each iron atom is homochiral.

The chiragen derivative **91**, incorporating two stereoselectively-linked bis-[4,5]-pineno-2,2'-bipyridine units, also yields complexes of type $[M_2L_3]^{4+}$ (M = Fe, Zn and Cd).[144]

From CD and NMR measurements, these were assigned triple-helical structures in which enantiomerically pure, homochiral configurations occur at the metal centres. In effect, the chirality associated with the pinene fragments appears to direct the assembly process such that only one isomer of the resultant dinuclear

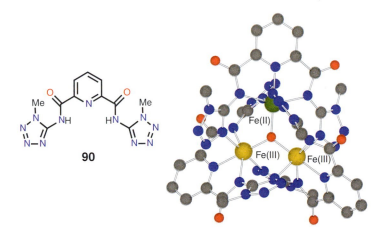

90

Figure 6.33 *The X-ray structure of the neutral species, $[Fe_3OL_3]$ (L = the doubly-deprotonated form of **90**)*[143]

142 D. Zurita, P. Baret and J.-L. Pierre, *New J. Chem.*, 1994, **18**, 1143.
143 R.W. Saalfrank, S. Trummer, H. Krautscheid, V. Schünemann, A.X. Trautwein, S. Hien, C. Stadler and J. Daub, *Angew. Chem., Int. Ed. Engl.*, 1996, **35**, 2206.
144 H. Mürner, A. von Zelewsky and G. Hopfgartner, *Inorg. Chim. Acta*, 1998, **271**, 36.

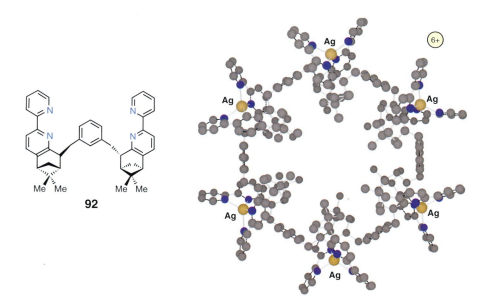

91

triple-helical structure is formed. Another example of the stereospecific self-assembly of a dinuclear iron(II) helicate of a bis(bipyridine)-containing ligand of the 'chiragen' family has also been demonstrated.[145]

An extension of the above study has produced a further example[146] of the expanding *cyclic* helicate category[76,147] mentioned previously. Interaction of **92** with silver(I) results in the spontaneous assembly of a circular, six-fold, single-helical species. The X-ray structure of this novel product is shown in Figure 6.34. The six silver ions are related by a crystallographic C_6-axis, with each being tetrahedrally-co-ordinated by four almost equidistant nitrogen donors located on two different

92

Figure 6.34 *X-Ray structure of the circular (six-fold), single-helical complex of* **92**[146]

[145] P. Baret, D. Gaude, G. Gellon and J.-L. Pierre, *New J. Chem.*, 1997, **21**, 1255.
[146] L. Mamula, A. von Zelewsky and G. Bernardinelli, *Angew. Chem., Int. Ed. Engl.*, 1998, **37**, 290.
[147] G. Hopfgartner, C. Piguet and J.D. Henion, *J. Am. Soc. Mass Spectrom.*, 1994, **5**, 748; P.L. Jones, K.J. Byrom, J.C. Jeffery, J.A. McCleverty and M.D. Ward, *Chem. Commun.*, 1997, 1361; P.N.W. Baxter, J.-M. Lehn and K. Rissanen, *Chem. Commun.*, 1997, 1323; C. Provent, S. Hewage, G. Brand, G. Bernardinelli, L.J. Charbonniere and A.F. Williams, *Angew. Chem., Int. Ed. Engl.*, 1997, **36**, 1287; D.P. Funeriu, J.-M Lehn, G. Baum and D. Fenske, *Chem. Eur. J.*, 1997, **3**, 99.

ligands. All chiral pinene groups are orientated towards the centre of the structure, thus giving rise to a 'chiral cavity' (with a diameter of 0.84 Å).

The earlier study by Shanzer *et al.*,[137] has formed the basis for the synthesis of a new class of molecular redox switch based on the chemical triggering of iron(II)/iron(III) translocation within triple-stranded helical complexes.[148] For example, the tripodal ligand **93** is capable of binding iron at either its 'terminal' tris-bipyridyl site or at the 'harder' site made up of three 'internal' hydroxamate binding groups (see Figure 6.35). Reversible, intramolecular translocation of the iron between these sites is able to be achieved by chemical oxidation and reduction – reflecting the different co-ordination preferences of iron(II) and iron(III). The softer iron(II) prefers the tris(bipyridine) site whereas the harder iron(III) prefers the tris(hydroxamate) site. The characteristic spectral properties of the respective complexes allow quantitative monitoring of the switching process. Thus, treatment

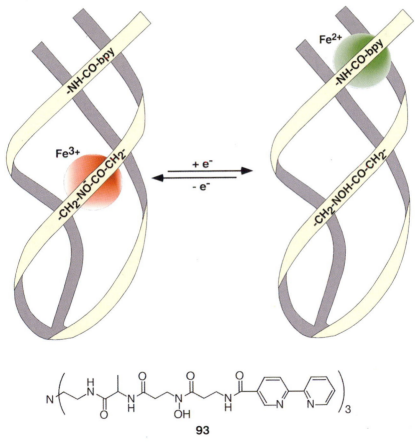

93

Figure 6.35 *The translation of iron in the triply-stranded helical complex of the tripodal ligand,* **93**[148]

[148]   L. Zellkovich, J. Libman and A. Shanzer, *Nature*, 1995, **374**, 790.

of the light-brown chiral complex containing iron(III) with ascorbic acid results in rapid reduction to the iron(II) analogue in which the chromophore is now the purple tris(bipyridyl)iron(II) derivative. The process can be reversed through oxidation by using ammonium persulfate at 70 °C.

6.6.2 Gallium and Titanium Systems

Related studies to the above leading to gallium(III), titanium(IV) or mixed gallium(III)–titanium(IV) dinuclear complexes have been reported.[149,150] For example, ligand **94** contains two different metal binding sites, and reaction with titanium(IV) or gallium(III) leads to complexes of type $[Ti_2L_3]^-$ and $[Ga_2L_3]^{3-}$.[150] When a 1 : 1

94

mixture of these ions was employed for the synthesis, then the mixed-metal species $[TiGaL_3]^{2-}$ was obtained. In each of the single-metal complexes, two ligands are orientated in one direction while the third co-ordinates in the opposite direction. In contrast, the mixed-metal complex contains all three ligand strands orientated in the same direction; the X-ray structure shows that the titanium(IV) binds at a tris(catecholate) site while the gallium(III) occupies a tris(aminophenolate) site. Both ions are in pseudo-octahedral environments with the metal centres having Λ and Δ configurations. Overall, the complex is non-helical. In part, the arrangement adopted appears to be a consequence of the use of methylene groups as spacers in the ligand system. It is noted that the stability of the superstructure in this complex is enhanced by hydrogen-bonding interactions between the NH₂ groups and 'internal' oxygen atoms belonging to the titanium(IV)-tris(catecholate) unit. This study serves to illustrate the manner by which different combinations of two metal ions can trigger the selective self-assembly of related (but structurally different) supramolecular aggregates.

Other studies involving the self-assembly of tri-stranded, non-chiral complexes composed of three bis(catecholate) ligands wrapped around two titanium(IV) ions have been reported.[151] The self-assembly of gallium(III) catecholamide triple helices has also been investigated[152,153] along with the assembly and racemisation of helicates incorporating linked catechol derivatives and either titanium(IV)[154] or gallium(III).[155]

[149] M. Albrecht, *Chem. Eur. J.*, 1997, **3**, 1466.
[150] M. Albrecht and R. Fröhlich, *J. Am. Chem. Soc.*, 1997, **119**, 1656.
[151] M. Albrecht and S. Kotila, *Angew. Chem., Int. Ed. Engl.*, 1995, **34**, 2134.
[152] E.J. Enemark and T.D.P. Stack, *Angew. Chem., Int. Ed. Engl.*, 1995, **34**, 996.
[153] D.L. Caulder and K.N. Raymond, *Angew. Chem., Int. Ed. Engl.*, 1997, **36**, 1440.
[154] M. Albrecht and M. Schneider, *Chem. Commun.*, 1998, 137 and references therein.
[155] B. Kersting, M. Meyer, R.E. Powers and K.N. Raymond, *J. Am. Chem. Soc.*, 1996, **118**, 7221.

In a further study,[152] the *trans*-influence (the *trans*-influence is a thermodynamic phenomenon that is concerned with the ground state effects of ligands that are orientated *trans* to one another, while the more widely known *trans*-effect is a kinetic phenomenon[156]) has been employed in the gallium(III)-assisted self-assembly of a dinuclear triple helix from ligands such as **95**. The latter only differ in the nature of the bridging propyl group present.

95

Once again, inherent in the design of these ligand species is their ability to co-ordinate to two metals at separated sites that for steric reasons are unable to bind simultaneously to a single metal ion.

As occurs for **95**, electronic asymmetry within the bidentate co-ordination domains may be used to provide an additional enthalpic drive towards the formation of a required dinuclear triple-helical species. Namely, stereochemical bias may arise from the operation of the above-mentioned *trans*-influence since there will be a tendency for stronger donor groups to orientate *trans* to donors with weaker donor properties. In a monomeric complex containing three similar bidentate ligands with donor groups of different 'strengths', it is only the facial configuration that has all the stronger donors opposite the weaker ones. Thus, it was reasoned that incorporation of two non-symmetric, bidentate subunits in each ligand strand (such as occurs in **95**) should promote structural specificity through co-operative binding behaviour in the formation of the 'usual' 2 : 3 (metal : ligand) triple helix.

The X-ray structure of the $[Ga_2L_3]^{6-}$ anion (where L = *R,R* form of **95**) is shown in Figure 6.36. Each gallium is surrounded by three catecholamide groups, with a metal separation of 10.78 Å – a long distance that presumably minimises repulsion between the three positive charges associated with each metal centre. Around each metal there are two sets of three Ga–O bond distances, consistent with the operation of the predicted *trans*-influence in this system. Both metal centres have the same absolute configuration (Λ, Λ), controlled by the absolute chirality of the enantiomerically pure diamine used for the synthesis of the ligand. Two lessons are apparent from this study: (i) that the *trans*-influence, a well characterised phenomenon in mononuclear systems, can be used through the medium of ligand design to enhance stereochemical specificity in oligonuclear systems; and (ii) that enantiomerically pure ligands can be successfully employed to produce a stereospecific helix, even when the ligand strands are relatively flexible.

---

[156] T.G. Appleton, H.C. Clark and L.E. Manzer, *Coord. Chem. Rev.*, 1973, **10**, 335.

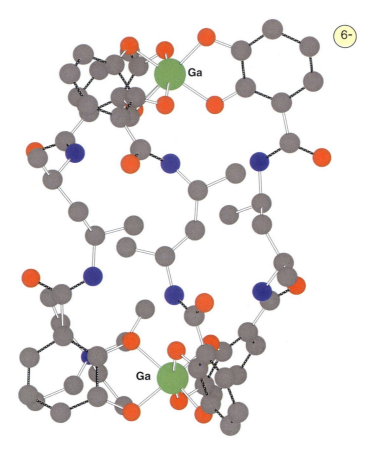

Figure 6.36 *X-Ray structure of [Ga$_2$L$_3$]$^{6-}$ (L = R,R form of **95**)*[152]

6.6.3   Lanthanide-containing Systems

Investigation of the structural and photophysical properties of selected triple-strand-ed lanthanide complexes incorporating planar tridentate (N$_3$-donor) ligands of types **96** and **97** have been reported.[157–160] Emphasis in such studies has been given to compounds based on europium(III) and terbium(III) since these ions give rise to efficient luminescent probes because of their long-lived excited states and their large Stokes shifts. Other requirements of importance in the design of efficient probes include:[161] (i) protection of the complexed lanthanide ion from the luminescence

[157]   G. Bernardinelli, C. Piguet and A.F. Williams, *Angew. Chem., Int. Ed. Engl.*, 1992, **31**, 1622.
[158]   C. Piguet, A.F. Williams, G. Bernardinelli and J.-C.G. Bünzli, *Inorg. Chem.*, 1993, **32**, 4139.
[159]   C. Piguet, J.-C.G. Bünzli, G. Bernardinelli, G. Hopfgartner and A.F. Williams, *J. Am. Chem. Soc.*, 1993, **115**, 8197.
[160]   C. Piguet, J.-C.G. Bünzli, G. Bernardinelli, G. Hopfgartner and A.F. Williams, *J. Alloys Compd.*, 1995, **225**, 324.
[161]   J.-C.G. Bünzli, P. Froidevaux and C. Piguet, *New J. Chem.*, 1995, **19**, 661.

96

97

quenching that may occur *via* interaction with solvent molecules and/or high-energy vibrations associated with the ligating groups; (ii) the presence of multiple absorbing groups that are suitable for energy transfer (the antenna effect); and (iii) sufficiently high thermodynamic and kinetic stabilities associated with the lanthanide ion complex.

The derivatives **96** (R^1, R^2 = H; R^3 = Me) and **97** were shown to undergo complexation with individual lanthanide(III) ions to yield stable mono-[158,161] and dinuclear[157,159,161] triple-stranded helicates. In these complexes, each lanthanide ion is surrounded by nine nitrogen donors in a pseudo-tricapped trigonal prismatic arrangement. The complex cation has pseudo D_3-symmetry, with the geometry being similar to that found in the corresponding [Eu(2,2′:6′2″-terpyridine)$_3$]$^{3+}$ cation.[162] It is noted that the effectiveness of this latter complex as a luminescent probe is somewhat reduced as a consequence of its tendency towards kinetic lability in solution (on average, one nitrogen donor site on the metal is replaced by a solvent molecule). In contrast, particular complexes of derivatives of type **96** have been demonstrated to be more kinetically robust.

The luminescent lanthanide (Eu, Gd or Tb) complexes of (**96** ; R^1 = H; R^2 = Me, R^3 = Et) provide further examples of mononuclear triple-helical complexes of the type just discussed.[163] The X-ray structure of the [EuL$_3$]$^{3+}$ cation shows that the three tridentate ligands are arranged around a pseudo-C_3 axis passing through the metal such that the co-ordination sphere of the europium is once again a (slightly) distorted trigonal-tricapped prism. In this, the six benzimidazole nitrogen atoms occupy the vertices of the prism, while the three pyridine nitrogens occupy capping positions.

Reaction of the segmented bidentate-tridentate ligand (**59**)[117] with zinc(II) and lanthanum(III) or europium(III) ions in acetonitrile yielded novel heteronuclear species of type [LnZn(L)$_3$]$^{5+}$. In part, the inherent selectivity associated with this procedure reflects the lanthanide ion's low affinity for the tridentate (six-co-ordinate)

[162] G.H. Frost, F.A. Hart, C. Heath and M.B. Hursthouse, *J. Chem. Soc., Chem. Commun.*, 1969, 1421; D.A. Durham, G.H. Frost and F.A. Hart, *J. Inorg. Nucl. Chem.*, 1969, **31**, 833.
[163] C. Piguet, J.-C.G. Bünzli, G. Bernardinelli, C.G. Bochet and P. Froidevaux, *J. Chem. Soc., Dalton Trans.*, 1995, 83.

binding site in the resulting (head-to-head) triple helix. As discussed above, it has been well demonstrated that ions of this type will prefer a pseudo-tricapped trigonal prismatic arrangement defined by three tridentate units – one from each of the respective ligand strands. On the other hand, six-co-ordinate zinc(II) in an N_6-donor environment is not an uncommon arrangement for this ion.

Other segmented ligand systems related to those just discussed have been employed for the preparation of further dinuclear, tri-stranded helicates incorporating both homo-, $[Ln_2L_3]^{6+}$, and hetero-, $[LnML_3]^{n+}$, metal ions.[164,165] The mixed-metal complexes include lanthanide(III)/iron(II)[164] and lanthanide(III)/cobalt(II)[165] systems. An example of the latter type was shown to be readily oxidised to the corresponding lanthanum(III)/cobalt(III) species. Owing to the considerable kinetic inertness of the bound cobalt(III) (d^6) ion, it has proved possible to remove the lanthanide ion selectively from this complex cation by treatment with a stoichiometric amount of $(NBu_4)EDTA$. This resulted in almost quantitative isolation of pure *fac*-$[CoL_3]^{3+}$ as its perchlorate salt. In the absence of the lanthanide 'templating' ion, a related procedure yielded the cobalt(III) product as a mixture of its meridional and facial forms.

6.7   Molecular Knots

Duplex circular DNA is known to be transformed into a number of catenated and knotted forms when treated with 'gyrase' enzymes.[166] Despite this parallel in Nature, the construction of synthetic molecules with knot topologies has long provided a challenge for chemists – especially in attempting to devise rational (non-random) syntheses for such structures.[167] This challenge has now been met, due largely to a number of elegant studies by Sauvage and his group.[53,168,169]

Using a synthetic strategy based on a metal template approach, Dietrich-Buchecker and Sauvage[22,170] were successful in synthesising a molecular trefoil knot (the simplest knot, consisting of three lobes and three crossing points) incorporating an 86-membered macrocycle containing 14 oxygen and 8 nitrogen atoms; the molecular weight of this supramolecular species is 1690. This impressive achievement rested heavily on the strategies developed earlier for the metal-directed interlocking of molecular threads.

[164]   C. Piguet, E. Rivara-Minten, G. Bernardinelli, J.-C.G. Bünzli and G. Hopfgartner, *J. Chem. Soc., Dalton Trans.*, 1997, 421.

[165]   S. Rigault, C. Piguet, G. Bernardinelli and G. Hopfgartner, *Angew. Chem., Int. Ed. Engl.*, 1998, **37**, 169 and references therein.

[166]   See, for example: B. Hudson and J. Vinograd, *Nature*, 1967, **216**, 647; D.A. Clayton and J. Vinograd, *Nature*, 1967, **216**, 652; J.C. Wang and H. Schwartz, *Biopolymers*, 1967, **5**, 953; L.F. Liu, R.E. Depew and J.C. Wang, *J. Mol. Biol.*, 1976, **106**, 439; K.N. Kreuzer and N.R. Cozzarelli, *Cell*, 1980, **20**, 245; M.A. Krasnow, A. Stasiak, S.J. Spengler, F. Dean, T. Koller and N.R. Cozzarelli, *Nature*, 1983, **304**, 559; S.A. Wasserman and N.R. Cozzarelli, *Proc. Natl. Acad. Sci. USA*, 1985, **82**, 1079; J.D. Griffith and H.A. Nash, *Proc. Natl. Acad. Sci. USA*, 1985, **82**, 3124; S.A. Wasserman and N.R. Cozzarelli, *Science*, 1986, **232**, 951 and references therein.

[167]   G. Schill, *Catenanes, Rotaxanes and Knots*, Academic Press, New York and London, 1971.

[168]   J.-P. Sauvage, *Acc. Chem. Res.*, 1990, **23**, 319; C. Dietrich-Buchecker and J.-P. Sauvage, *Bull. Soc. Chem. Fr.*, 1992, **129**, 113.

[169]   C. Dietrich-Buchecker and J.-P. Sauvage, *New J. Chem.*, 1992, **16**, 277.

[170]   C.O. Dietrich-Buchecker and J.-P. Sauvage, *Angew. Chem., Int. Ed. Engl.*, 1989, **28**, 189.

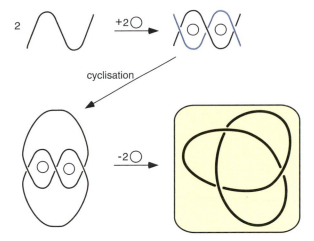

Figure 6.37 *Schematic representation of a strategy for preparing a trefoil knot*[170]

It needs to be noted that a trefoil knot is only the first member of a potential series of related molecular knots.[169] If the number of metal centres employed in generating the associated helix is even, then the number of crossing points will be odd, and a single, closed knotted structure will be formed (trefoil, pentafoil, heptafoil, *etc.*). On the other hand, if the number of metal centres is odd, then an even number of crossing points will result, giving rise to multiply-interlocked [2]-catenanes.

The approach employed for the preparation of the trefoil knot just mentioned is outlined schematically in Figure 6.37. It proceeds *via* a dinuclear double-helical intermediate. The following steps were followed: (i) two ligand strands were co-ordinated to two metal centres in a normal metal template reaction to produce a double-helical precursor; (ii) cyclisation was then carried out by linking the ends of each ligand strand in the manner shown, and (iii) this was followed by demetallation to yield the required metal-free trefoil knot. Essential to the success of this procedure is the need for the intermediate dicopper(I) species to be of sufficient stability to survive the rigours of the cyclisation reaction. In view of this, the use of 1,10-phenanthroline units in the ligand strands is an appropriate choice because of their well known high affinity for copper(I) (as mentioned elsewhere in this discussion). Similarly, the use of appropriate spacers between the phenanthroline domains, as well as the choice of copper(I) for the templating metal, were both design features arising from experience gained in the prior studies.

The double-helical, dinuclear copper(I) complex **98** was isolated as a deep red solid by addition of a stoichiometric amount of [Cu(CH$_3$CN)$_4$](BF$_4$) in acetonitrile to a solution of the corresponding free ligand in *N,N*-dimethylformamide. Reaction of this complex with two equivalents of diiodo-hexaethyleneglycol in the presence of a large excess of Cs$_2$CO$_3$ led to a mixture of products from which it proved possible to isolate the dimetalled knot **99** in 3% yield. Quantitative demetallation of the various copper(I) species formed in the reaction mixture also led to the isolation of small amounts of each of the metal-free products **100**, **101** and **102**. In contrast to the isomeric large ring macrocycle **101** which shows a relatively well-resolved [1]H

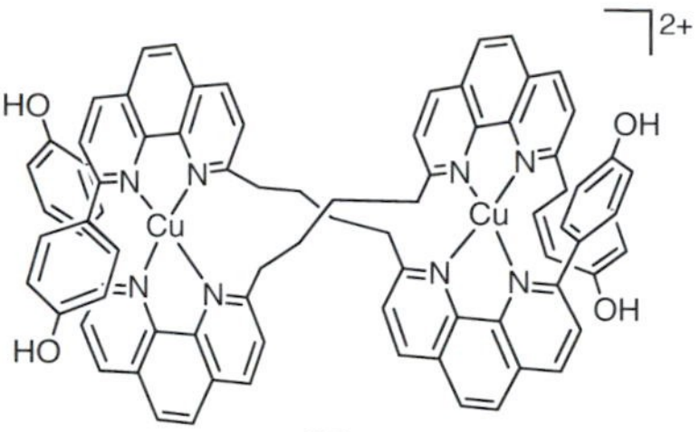

98

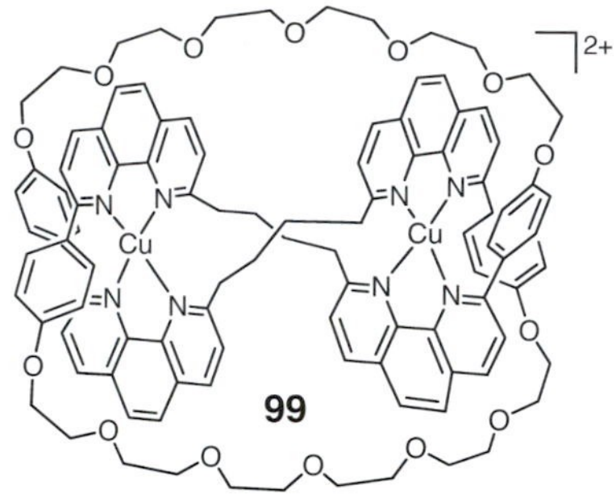

99

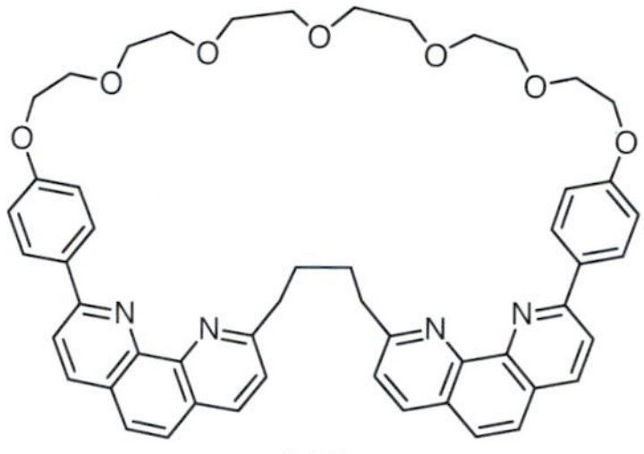

100

101

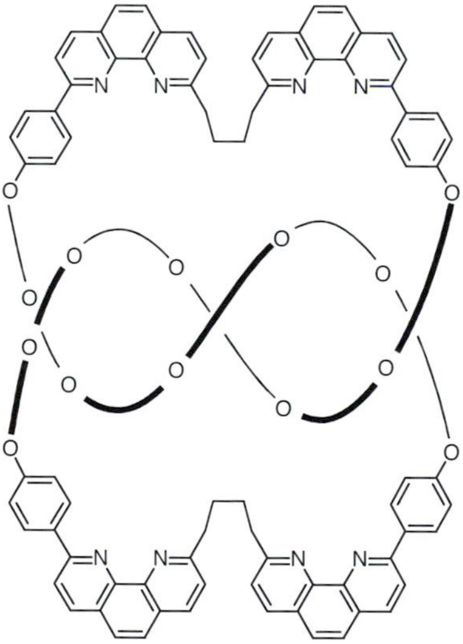

102

NMR spectrum, a distinctive feature of the spectrum of the trefoil structure **102** is the presence of broad, poorly-resolved peaks in the aromatic region.

The initial structural assignment of the above trefoil knot was based on mass spectral as well as ^{1}H NMR data. However, its structure was subsequently confirmed by an X-ray diffraction study of the dinuclear copper(I) precursor (see Figure 6.38).[171] The two copper(I) ions are located 6.3 Å apart inside the double helix.

[171] C.O. Dietrich-Buchecker, J. Guilhem, C. Pascard and J.-P. Sauvage, *Angew. Chem., Int. Ed. Engl.*, 1990, **29**, 1154.

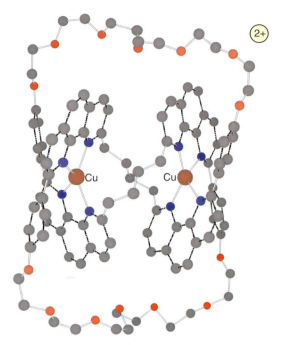

Figure 6.38 *Structure of the double-helical, dinuclear copper(I) complex,* **99**[22,170]

Variation of the structural parameters of the respective components employed in the construction of a series of related trefoil knots resulted in some improvement in the yields of individual products.[172] These studies led to both a range of new dicopper(I)-containing knots as well as to a series of corresponding face-to-face isomers. In the latter species no crossing of the strands occurs (see Figure 6.39). Both the knots and the face-to-face isomers contained 80- to 90-membered macroyclic rings, with the respective yields being very dependent upon such parameters as the number of methylene groups acting as spacers between the co-ordination domains. For the knots, the highest yield (8%) was obtained for an 84-membered ring system incorporating $-(CH_2)_6-$ spacers. In a number of instances, the dicopper(I) face-to-face isomers were obtained as the major products; for one face-to-face complex, a yield of 24% was obtained.

In a further study, a 1,3-phenylene linker was employed as the bridge between two 1,10-phenanthroline fragments in the ligand strand (see **103**). The corresponding dicopper(I) double-helical complex was found to contain the two ligands tightly wound around the copper centres (the Cu–Cu distance is very short at 4.76 Å) and, as a result, this complex was judged to be a suitable precursor for formation of the corresponding trefoil knot.[173] Reaction of the appropriate ends of the

[172] C.O. Dietrich-Buchecker, J.-F. Nierengarten and J.-P. Sauvage, *Pure Appl. Chem.*, 1994, **66**, 1543; C.O. Dietrich-Buchecker, J.-F. Nierengarten, J.-P. Sauvage, N. Armaroli, V. Balzani and L. De Cola, *J. Am. Chem. Soc.*, 1993, **115**, 11237.
[173] C.O. Dietrich-Buchecker, J.-P. Sauvage, A. De Cian and J. Fischer, *J. Chem. Soc., Chem. Commun.*, 1994, 2231.

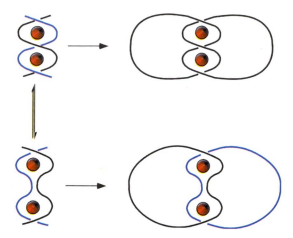

Figure 6.39 *Preparation of a dimetallic 'trefoil' complex and its corresponding 'face-to-face' isomer[172]*

103

complexed strands with $ICH_2(CH_2OCH_2)_5CH_2I$ and Cs_2CO_3 in dimethylformamide at 60 °C gave the required trefoil product in 29% yield (after chromatography). Reflecting its 'tight' structure, demetallation experiments (including kinetic and mechanistic studies)[174] indicated that this product is significantly inert towards copper(I) decomplexation in the presence of cyanide ion, relative to both the analogous alkyl-linked knots as well as the related face-to-face isomers.

A trefoil knot is a chiral object. The above 1,3-phenylene-linked dicationic knot has been resolved *via* fractional crystallisation of its diastereomeric salts using the chiral anion (*S*)-(+)-1,1′-binaphthyl-2,2′-diyl phosphate.[175]

Related to its overall kinetic inertness, the above dicopper(I) complex of the 1,3-phenylene-containing knot was able to be demetallated in a two-step sequence.[176] Such behaviour has been used synthetically to produce hetero-binuclear species.

[174] A.M. Albrecht-Gary, C.O. Dietrich-Buchecker, J. Guilhem, M. Meyer, C. Pascard and J.-P. Sauvage, *Recl. Trav. Chim. Pays-Bas*, 1993, **112**, 427; M. Meyer, A.-M. Albrecht-Gary, C.O. Dietrich-Buchecker and J.-P. Sauvage, *J. Am. Chem. Soc.*, 1997, **119**, 4599.

[175] G. Rapenne, C. Dietrich-Buchecker and J.-P. Sauvage, *J. Am. Chem. Soc.*, 1996, **118**, 10932.

[176] C.O. Dietrich-Buchecker, J.-P. Sauvage, N. Armaroli, P. Ceroni and V. Balzani, *Angew. Chem. Int., Ed. Engl.*, 1996, **35**, 1119.

Thus, the vacant co-ordination site in the singly-demetallated product is readily re-metallated with zinc(II) or silver(I) to yield species of type $[CuZnL]^{3+}$ and $[CuAgL]^{2+}$.

6.7.1 A High Yield Trefoil Synthesis

A high yield synthesis for the dicopper(I) trefoil knot of type **104** has been reported.[177] The overall reaction sequence is shown in Figure 6.40. In this procedure, two metal centres act as the template for the assembly of a double-stranded helix bearing four terminal alkene functional groups. This 'pendant arm' species is then converted into the corresponding copper-containing trefoil knot in 74% yield by means of two ruthenium(II)-catalysed, ring-closing reactions. The final copper-containing, 82-membered knot **104** is obtained by catalytic hydrogenation of the (cyclic) alkene functions formed during the ring-closing reaction.

The use of the metal ion template procedure for producing topologically sophisticated structures is well illustrated by the synthesis of 'composite' molecular knots such as **105** and **106** (Figure 6.41); in this case two topological diasteromers are

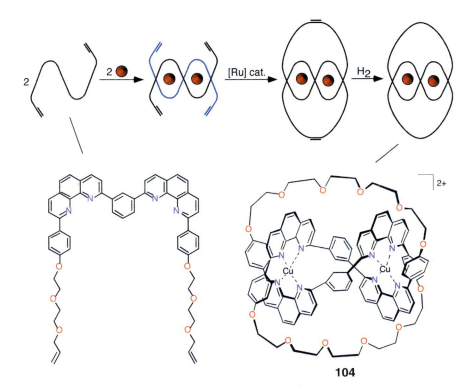

104

Figure 6.40 *High yield synthesis of the cationic 'trefoil' complex, **104**, from its open-chain precursor and copper(I)*[177]

177   C. Dietrich-Buchecker, G. Rapenne and J.-P. Sauvage, *Chem. Commun.*, 1997, 2053.

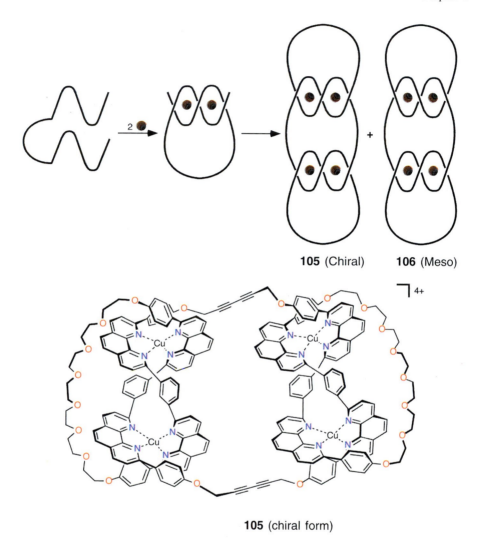

105 (Chiral)   **106** (Meso)

105 (chiral form)

Figure 6.41 *Formation of 'composite' molecular knots*[178]

formed.[178] Other related composite knots are also produced in this reaction. These products, shown in the figure, are composite knots rather than prime knots (the tre-foil knot is an example of a prime knot) since two areas of intertwining are present in their structure. The yields of these products were quite low (at around 3%), undoubtedly reflecting the number of 'crossings' that occurs in their construction.

[178] R.C. Carina, C. Dietrich-Buchecker and J.-P. Sauvage, *J. Am. Chem. Soc.,* 1996, **118**, 9110.

CHAPTER 7

Further Metal-containing Systems

7.1 Introduction

Besides the metal-containing categories discussed in the last chapter, the world of supramolecular architecture contains a very large number of other self-assembled systems in which metal ions play a role as structural elements. Included among these are metallo-rings, squares, grids, ladders, boxes, rods, tubes and a number of systems having less readily defined shapes. Many of these species contain voids and hence form a group of (real or potential) artificial receptors, often categorised by being formed by metal-templated self-assembly.[1,2] In this chapter the formation and properties of a selection of examples from the above structural types are discussed.

As in many of the metal-containing assemblies discussed in Chapter 6, a considerable number of other systems incorporating 'classical' organic ligand fragments have been synthesised. For example, systems containing bipyridyl or terpyridyl entities are common. Ligand species of the latter type remain popular choices since they can act both as good σ- and π-donors as well as π-acceptors and will hence tend to stabilise both high and low oxidation states of a bound metal ion. Such ligands also tend to yield transition metal complexes that exhibit substantial thermodynamic stability, while themselves being chemically resistant to degradation under a variety of conditions. Although a wide range of metal ions has been employed in the various studies, once again, frequent use has been made of copper(I) (tetrahedral), silver(I) (tetrahedral) and ruthenium(II) (octahedral) as structural elements. As well as acting as centres for ligand binding and orientation, as already mentioned elsewhere, the presence of metal ions in many cases introduces electrochemical and/or photochemical properties that confer additional functionality to the final supramolecular structures.

[1]   J.R. Fredericks and A.D. Hamilton, *Perspect. Supramol. Chem.*, 1996, **3**, 1; T.J. Hubin, A.G. Kolchinski, A.L. Vance and D.H. Busch, *Adv. Supramolecular Chem.*, 1999, **15**, 237.
[2]   C.J. Jones, *Chem. Soc. Rev.*, 1998, **27**, 289; *Transition Metals in Supramolecular Chemistry*, ed. J.P. Sauvage, Wiley, New York, 1999.

7.2 Metallocycles – Rings, Squares and Related Species

7.2.1 Rings

A considerable number of large ring assemblies incorporating metal atoms as part of their ring structures are now known.[3]

The self-assembled metallocycle **2** represents a simple (dinuclear) example of the above type. This species forms spontaneously on mixing solutions of the diphosphine **1** and sodium tetrachloropalladate(II).[4] The success of this process, which appears to proceed under thermodynamic control (since a high-dilution procedure is not required), clearly depends upon a number of influences. These include the relatively high affinity of the phosphorus donors for the soft palladium(II) ion and the use of a semi-rigid bis(methylphenyl)ether linkage between the phosphorus donors. The latter inhibits simple bidentate co-ordination of both phosphorus atoms of **2** to a single palladium but is still flexible enough to allow bending about the ether linkage such that an ellipsoidal dinuclear geometry can be achieved.

With respect to the above, it has been recognised that enthalpy will tend to favour formation of a cyclic species over a corresponding linear product.[5,6] This is because

3   See, for example: (a) A.J. Pryde, B.L. Shaw and B. Weeks, *J. Chem. Soc., Chem. Commun.*, 1973, 947; (b) F.C. March, R. Mason, K.M. Thomas and B.L. Shaw, *J. Chem. Soc., Chem. Commun.*, 1975, 584; (c) A.R. Sanger, *J. Chem. Soc., Chem. Commun.*, 1975, 893; (d) N.A. Al-Salem, H.D. Empsall, R. Markham, B.L. Shaw and B. Weeks, *J. Chem. Soc., Dalton Trans.*, 1979, 1972; (e) R.J. Puddephatt, *Chem. Soc. Rev.*, 1983, **12**, 99; (f) A.W. Maverick and F.E. Klavetter, *Inorg. Chem.*, 1984, **23**, 4129; (g) A.W. Maverick, S.C. Buckingham, Q. Yao, J.R. Bradbury and G.G. Stanley, *J. Am. Chem. Soc.*, 1986, **108**, 7430; (h) W.E. Hill, J.G. Taylor, C.P. Falshaw, T.J. King, B. Beagley, D.M. Tonge, R.G. Pritchard and C.A. McAuliffe, *J. Chem. Soc., Dalton Trans.*, 1986, 2289; (i) J.R. Bradbury, J.L. Hampton, D.P. Martone and A.W. Maverick, *Inorg. Chem.*, 1989, **28**, 2392; (j) A.W. Maverick, M.L. Ivie, J.H. Waggenspack and F.R. Fronczek, *Inorg. Chem.*, 1990, **29**, 2403; (k) Y. Kobuke, Y. Sumida, M. Hayaswhi and H. Ogoshi, *Angew. Chem., Int. Ed. Engl.*, 1991, **30**, 1496; (l) H.-J. Schneider and D. Ruf, *Angew. Chem., Int. Ed. Engl.*, 1990, **29**, 1159; (m) S.I. Al-Resayes, P.B. Hitchcock and J.F. Nixon, *J. Chem. Soc., Chem. Commun.*, 1991, 78; (n) P. Scrimin, P. Tecilla, U. Tonellato and N. Vignaga, *J. Chem. Soc., Chem. Commun.*, 1991, 449; (o) E.C. Constable, R.P.G. Henney, P.R. Raithby and L.R. Sousa, *Angew. Chem., Int. Ed. Engl.*, 1991, **30**, 1363; (p) L.G. Mackay, H.L. Anderson and J.K.M. Sanders, *J. Chem. Soc., Chem. Commun.*, 1992, 43; (q) Y. Kobuke and Y. Satoh, *J. Am. Chem. Soc.*, 1992, **114**, 789; (r) A.W. Schwabacher, J. Lee and H. Lei, *J. Am. Chem. Soc.*, 1992, **114**, 7597; (s) K.L. Cole, M.A. Farran and K. Deshayes, *Tetrahedron Lett.*, 1992, **33**, 599; (t) S. Rüttimann, G. Bernardinelli and A.F. Williams, *Angew. Chem., Int. Ed. Engl.*, 1993, **32**, 392; (u) J. Lee and A.W Schwabacher, *J. Am. Chem. Soc.*, 1994, **116**, 8382; (v) M. Albrecht and S. Kotila, *Angew. Chem., Int. Ed. Engl.*, 1995, **34**, 2134; (w) M. Albrecht and S. Kotila, *Chem. Commun.*, 1996, 2309; (x) F.M. Romero, R. Ziessel, A. Dupont-Gervais and A. Van Dorsselaer, *Chem. Commun.*, 1996, 551; (y) F. Heirtzler and T. Weyhermüller, *J. Chem. Soc., Dalton Trans.*, 1997, 3653; (z) J.-I. Setsune, S. Muraoka and T. Yokoyama, *Inorg. Chem.*, 1997, **36**, 5135; (aa) E.C. Constable and E. Schofield, *Chem. Commun.*, 1998, 403; (ab) I.O. Fritsky, H. Kozlowski, E.V. Prisyazhnaya, A. Karaczyn, V.A. Kalibabchuk and T. Glowiak, *J. Chem. Soc., Dalton Trans.*, 1998, 1535; (ac) M.J. Hannon, C.L. Painting and W. Errington, *Chem. Commun.*, 1998, 307; (ad) R. Schneider, M.W. Hosseini, J.M. Planeix, A. De Cian and J. Fischer, *Chem. Commun.*, 1998, 1625; (ae) A.K. Duhme, S.C. Davies and D.L. Hughes, *Inorg. Chem.*, 1998, **37**, 5380; (af) R.W. Saalfrank, V. Seitz, D.L. Caulder, K.N. Raymond, M. Teichert and D. Stalke, *Eur. J. Inorg. Chem.*, 1998, 1313.
4   M. Fujita, J. Yazaki, T. Kuramochi and K. Ogura, *Bull. Chem. Soc. Jpn.*, 1993, **66**, 1837.
5   X. Chi, A.J. Guerin, R.A. Haycock, C.A. Hunter and L.D. Sarson, *J. Chem. Soc., Chem. Commun.*,1995, 2563.
6   P.J. Stang, N.E. Persky and J. Manna, *J. Am. Chem. Soc.*, 1997, **119**, 4777.

of the larger number of favourable interactions (namely, metal-donor bonds in the above example) per unit building block associated with formation of the cyclic system relative to formation of the corresponding linear oligomer composed of the same number of subunits. Furthermore, there is expected to be a favourable entropy contribution towards formation of a cyclic product involving the minimum number of subunits. This arises since, overall, the cost in terms of 'loss of degrees of freedom' on moving from the components to the product will be less for formation of a system in which the number of subunits used is at a minimum. Of course, as always, the formation of the thermodynamic product may be inhibited in a given preparation by kinetic barriers or by low solubility of an intermediate, causing the latter to preferentially precipitate from solution. Furthermore, the assembly process may not proceed to completion resulting in the formation of an equilibrium mixture of cyclic and non-cyclic (including precursor) species in solution.

Other metallocycles assembled from two ligand molecules and two metal atoms have been reported by Harding *et al.*[7–9] Bimetallic [2+2] macrocycles from the bis(bipyridyl) derivatives **3–7** and nickel(II), copper(I), silver(I), zinc(II) and cadmium(II) have been reported. Products of this type may have either chiral, helical

7   A. Bilyk and M.M. Harding, *J. Chem. Soc., Dalton Trans.*, 1994, 77.
8   A. Bilyk, M.M. Harding, P. Turner and T.W. Hambley, *J. Chem. Soc., Dalton Trans.*, 1994, 2783.
9   A. Bilyk, M.M. Harding, P. Turner and T.W. Hambley, *J. Chem. Soc., Dalton Trans.*, 1995, 2549.

or achiral, non-helical structures; the nature of the complex(es) formed being dependent on both the metal ion and the type of 'spacer' group present in the ligand. For example, the self-assembly of a bimetallic [2+2] metallocycle from two zinc(II) ions and two bis(bipyridyl) moieties linked by a 2,7-disubstituted pyrene group (see Figure 7.1) occurs spontaneously in solution. This product has been isolated in the solid state and its structure confirmed by X-ray diffraction.[8] The zinc ions are each bound to two bipyridyl groups and two oxygens from the ether links such that each metal has an octahedral co-ordination geometry. The overall arrangement is non-helical with the pyrene rings being offset by 3.84 Å and separated by 3.65 Å. The Zn···Zn distance is 14.42 Å. There is little evidence of significant π–π stacking between pyrene rings in this system. Interestingly, when the pyrene moieties are replaced by 1,4-disubstituted benzene groups 6, then approximately equimolar quantities of helical and non-helical (zinc) complexes form in equilibrium.

An extension of this work has resulted in the synthesis of related 6,6-disubstituted (bipyridyl) ligand derivatives containing pyromellitimide spacers connected with propynyl, butynyl and propynyloxy linkers.[10] The first of these ligands was shown to yield a chiral di-zinc(II) cationic complex of type $[Zn_2L_2]^{4+}$. This species is stabilised by the presence of an aromatic guest (p- or o-dimethoxybenzene) in its cavity. Related dinuclear species were also shown to form with cadmium(II) and nickel(II).

Many trinuclear metallocyclic species are known.[3t,11] Amongst these, examples

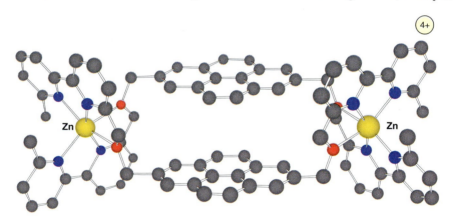

Figure 7.1 *Strucure of di-zinc(II) metallocycle incorporating bis(bipyridyl) moieties linked by 2,7-substituted pyrene groups*[8]

[10]   M.A. Houghton, A. Bilyk, M.M. Harding, P. Turner and T.W. Hambley, *J. Chem. Soc., Dalton Trans.*, 1997, 2725.
[11]   See, for example: (a) W. Burger and J. Strahle, *Z. Anorg. Allg. Chem.*, 1985, **529**, 111; (b) H.H. Murray, R.G. Raptis and J.P. Fackler, *Inorg. Chem.*, 1988, **27**, 26; (c) R.G. Raptis and J.P. Fackler, *Inorg. Chem.*, 1990, **29**, 5003; (d) H. Chen, M.M. Olmstead, D.P. Smith, M.F. Maestre and R.H. Fish, *Angew. Chem., Int. Ed. Engl.*, 1995, **34**, 1514; (e) R.W. Saalfrank, S. Trummer, H. Krautscheid, V. Schunemann, A.X. Trautwein, S. Hien, C. Stadler and J. Daub, *Angew. Chem., Int. Ed. Engl.*, 1996, **35**, 2206; (f) L. Schenetti, G. Bandoli, A. Dolmella, G. Trovo and B. Longato, *Inorg. Chem.*, 1994, **33**, 3169; (g) K.Yamanari, I. Fukuda, T. Kawamoto, Y. Kushi, A. Fuyuhiro, N. Kubota, T. Fukuo and R. Arakawa, *Inorg. Chem.*, 1998, **37**, 5611.

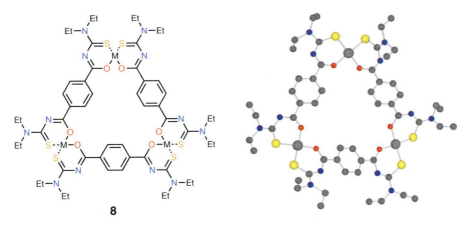

8

Figure 7.2 *The structure of the 27-membered metallocycle* **8**[13]

that incorporate 'linear', or almost linear, aromatic spacers between their metal centres (to define a triangular shape) are less common.[12] Compounds of this latter type include the nickel(II) and copper(II) trimeric species **8** (Figure 7.2), which incorporate 27-membered rings (counting the 'internal' oxygens in the ring).[13] Figure 7.2 also shows the X-ray structure of the nickel derivative.

7.2.2 Squares

The self-assembly of cyclophane-like square structures, comprised of *cis* bridging metal centres connected by rigid aromatic sides, represents the formation of another category of metallocycles which typically enclose cavities of nanometre dimensions. Although isolated structures of this type have been known for some time,[14,15] it is only relatively recently that increasing attention has been given to the area.

In general, the construction of a square geometry requires the assembly of rigid 'sides', held together by corner units, which incorporate valencies directed at 90°. Apart from the synthetic challenges inherent in producing such a geometry, the generation of a central cavity – of potential use in catalysis, separation technology and for sensing – has provided an additional motivation for undertaking investigations in the area.[16,17]

Tetranuclear species showing square geometries are generated spontaneously on reaction of dinitro(ethane-1,2-diamine)palladium(II) or its platinum(II) analogue

12   R.-D. Schnebeck, L. Randaccio, E. Zangrado and B. Lippert, *Angew. Chem., Int. Ed. Engl.*, 1998, **37**, 119.
13   R. Köhler, R. Kirmse, R. Richter, J. Sieler, E. Hoyer and L. Beyer, *Z. Anorg. Allg. Chem.*, 1986, **537**, 133.
14   W.C. Kalb, Z. Demidowicz, D.M. Speckmann, C. Knobler, R.G. Teller and M.F. Hawthorne, *Inorg. Chem.*, 1982, **21**, 4027.
15   P.M. Stricklen, E.J. Volcko and J.G. Verkade, *J. Am. Chem. Soc.*, 1983, **105**, 2494.
16   M. Fujita and K. Ogura, *Coord. Chem. Rev.*, 1996, **148**, 249.
17   M. Fujita and K. Ogura, *Bull. Chem. Soc. Jpn.*, 1996, **69**, 1471.

with 4,4'-bipyridine.[18,19] Integral to the success of this procedure is the use of square-planar complexes exhibiting a *cis* geometry such that two adjacent co-ordination sites (those occupied by the less strongly bound nitro groups) are available for binding to the incoming pyridyl nitrogen donors.

While the formation of the palladium species was shown to be fast (and in better than 90% yield), the kinetically less labile platinum(II) ion initially gave a mixture of oligomers; however, these kinetic products slowly rearranged (over a month) to the required thermodynamic product **9**. Models predict that free rotation of the 4,4'-bipyridine units is restricted such that they will prefer to be orientated perpendicular to the plane of the square; such an arrangement has been confirmed for the above tetra-palladium structure.[20]

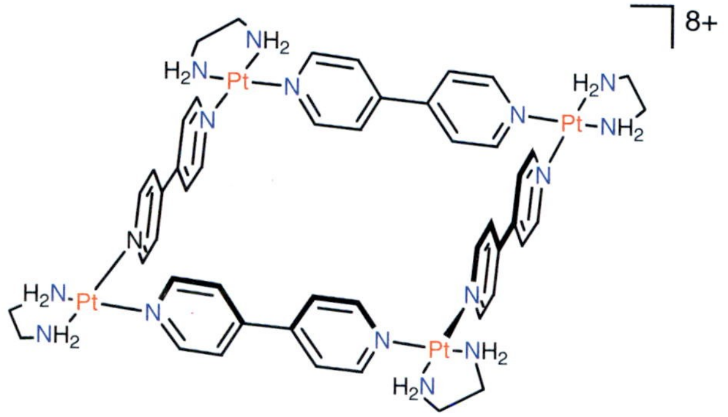

9

The use of longer chain (fully conjugated) bipyridine derivatives in experiments of the above type produced the corresponding analogues of **9** of type **10–13** incor-

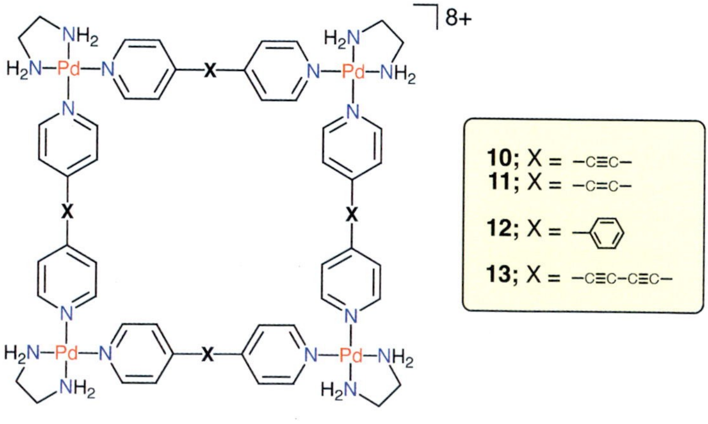

10; X = –C≡C–
11; X = –C=C–
12; X = –⟨◯⟩–
13; X = –C≡C–C≡C–

18   M. Fujita, J. Yazaki and K. Ogura, *J. Am. Chem. Soc.*, 1990, **112**, 5645.
19   M. Fujita, J. Yazaki and K. Ogura, *Chem. Lett.*, 1991, 1031.
20   M. Fujita, O. Sasaki, T. Mitsuhashi, T. Fujita, J. Yazaki, K. Yamaguchi and K. Ogura, *Chem. Commun.*, 1996, 1535.

porating larger voids.[16,20,21] However, the required product in all cases was accompanied by the formation of a by-product that was assigned to the corresponding cyclic trimer. The trimers will tend to be favoured entropically because they assemble from fewer components – as mentioned earlier.

The palladium(II)-containing assembly of type **10** and its platinum(II) analogue have been demonstrated to bind a range of electron-rich aromatic guests in their cavities. In part, such complexation is a reflection of charge-transfer interactions as well as being due to a contribution from a hydrophobic interaction between the electron-deficient pyridyl moieties (the electron deficiency is aided by co-ordination to the divalent metal) and the electron-rich aromatic guests. It has been demonstrated that the above self-assembled species of palladium(II) binds a 1,3,5-trimethoxybenzene guest in its central cavity in aqueous media with an association constant of 7.5×10^2 dm^3 mol^{-1}.[18,22]

When cadmium nitrate was treated with 4,4′-bipyridine (1 : 2 molar ratio) in a water–ethanol mixture, then colourless crystals of a corresponding 'infinite' complex species were obtained.[23] Thus, in the absence of *cis*-blocking ligands on the metal, as might be predicted, a polymeric array is formed.

The fact that iodonium groups are T-shaped with 90° angles, led Stang and Zhdankin[24] to employ such groups for the construction of the novel square **14** in which biphenyls form the sides. However, this iodonium-based square was not obtained by spontaneous self-assembly but is the product of a classical (stepwise) covalent synthesis.

14

The self-assembly strategy discussed previously was also employed by the Stang group to prepare rigid tetranucleated cationic squares, **15** and **16**, in which the corners were once again occupied by palladium(II) or platinum(II) ions in square planar environments – see Figure 7.3.[25] The precursor di-triflato complexes were employed since the triflato ligand is readily lost from the respective co-ordination spheres in the presence of a stronger binding group (in this case, one end of a

21 S.B. Lee, S. Hwang, D.S. Chung, H. Yun and J.-I. Hong, *Tetrahedron Lett.*, 1998, **39**, 873.
22 M. Fujita, J. Yazaki and K. Ogura, *Tetrahedron Lett.*, 1991, **40**, 5589.
23 M. Fujita, Y.J. Kwon, S. Washizu and K. Ogura, *J. Am. Chem. Soc.*, 1994, **116**, 1151.
24 P.J. Stang and V.V. Zhdankin, *J. Am. Chem. Soc.*, 1993, **115**, 9808.
25 P.J. Stang and D.H. Cao, *J. Am. Chem. Soc.*, 1994, **116**, 4981.

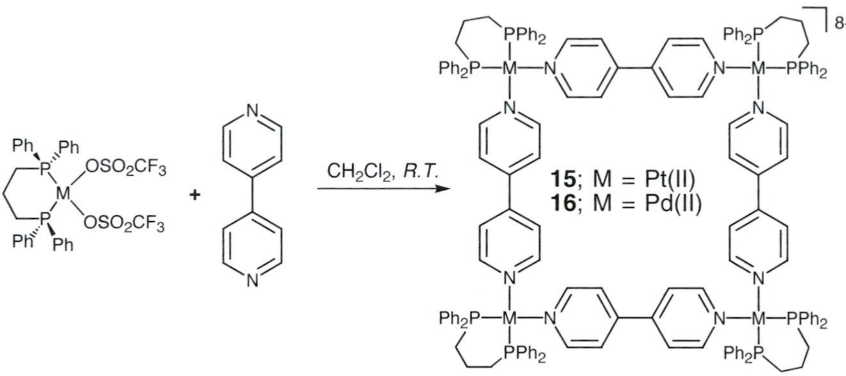

Figure 7.3 *Synthesis of the tetranucleated cationic squares* **15** *and* **16**[25]

4,4′-bipyridine ligand). The final self-assembled squares were found to be both soluble in a number of common organic solvents as well as being chemically robust.

The generality of the above reaction type has been demonstrated by employing both a range of other corner groups (that included chiral-ligand metal derivatives) and rigid (largely conjugated aromatic) sides to produce a wide variety of molecular squares.[26,27] Included in these are mixed-metal systems incorporating palladium(II) and platinum(II) (in which each side unit is non-symmetrical), as well as hybrid iodonium–palladium or iodonium–platinum systems and 'chiral' systems. A selection of products is given by **17–20**.[28–30]

Most of the assemblies produced in the above studies tend to be near-perfect squares, although the hybrid systems show rhomboidal distortions. These respective structures are relatively rigid; since they are cations (mainly with 8+ charges), they also show considerable potential as anion receptors. X-Ray studies indicate that in some cases,[27,28] these composite molecules stack in the solid state to produce channels and hence such species are of potential interest as artificial 'zeolites'.

Paralleling an earlier study,[15] new molecular squares have been constructed using octahedral rhenium carbonyl derivatives as the corner units. The resulting products are photoluminescent – a potentially useful property for use in sensing applications that involve inclusion of an appropriate guest molecule into the cavity of the

[26] C.A. Hunter and L.D. Sarson, *Angew. Chem., Int. Ed. Engl.*, 1994, **33**, 2313; P.J. Stang and J.A. Whiteford, *Organometallics*, 1994, **13**, 3776; P.J. Stang and K. Chen, *J. Am. Chem. Soc.*, 1995, **117**, 1667; P.J. Stang and J.A. Whiteford, *Res. Chem. Intermed.*, 1996, **22**, 659; P.J. Stang, B. Olenyuk, J. Fan and A.M. Arif, *Organometallics*, 1996, **15**, 904; P.J. Stang and N.E. Persky, *Chem. Commun.*, 1997, 77; P.J. Stang, J. Fan and B. Olenyuk, *Chem. Commun.*, 1997, 1453; P.J. Stang, D.H. Cao, K. Chen, G.M. Gray, D.C. Muddiman and R.D. Smith, *J. Am. Chem. Soc.*, 1997, **119**, 5163; K. Funatsu, T. Imamura, A. Ichimura and Y. Sasaki, *Inorg. Chem.*, 1998, **37**, 1798; R.V. Slone, K.D. Benkstein, S. Belanger, J.T. Hupp, I.A. Guzei and A.L. Rheingold, *Coord. Chem. Rev.*, 1998, **171**, 221; C. Müller, J.A. Whiteford and P.J. Stang, *J. Am. Chem. Soc.*, 1998, **120**, 9827.
[27] P J. Stang, K. Chen and A.M. Arif, *J. Am. Chem. Soc.*, 1995, **117**, 8793.
[28] P.J. Stang, D.H. Cao, S. Saito and A.M. Arif, *J. Am. Chem. Soc.*, 1995, **117**, 6273.
[29] B. Olenyuk, J.A. Whiteford and P.J. Stang, *J. Am. Chem. Soc.*, 1996, **118**, 8221.
[30] J. Manna, J.A. Whiteford, P.J. Stang, D.C. Muddiman and R.D. Smith, *J. Am. Chem. Soc.*, 1996, **118**, 8731.

square.[31] The related rhenium/palladium(II) square **21** is also visible-light-address-able; however, the presence of the palladium(II) ions attenuates the degree of pho-toluminescence in this case.[32]

An innovative extension of the general strategy for producing molecular squares has resulted in the formation of discrete, cavity-containing rectangular grids of the type illustrated in Figure 7.4.[33] These products were obtained by the one-pot

[31] R.V. Slone, J.T. Hupp, C.L. Stern and T.E. Albrecht-Schmitt, *Inorg. Chem.*, 1996, **35**, 4096.
[32] R.V. Slone, D.I. Yoon, R.M. Calhoun and J.T. Hupp, *J. Am. Chem. Soc.*, 1995, **117**, 11813.
[33] L.F. MacGillivray, R.H. Groeneman and J.L. Atwood, *J. Am. Chem. Soc.*, 1998, **120**, 2676.

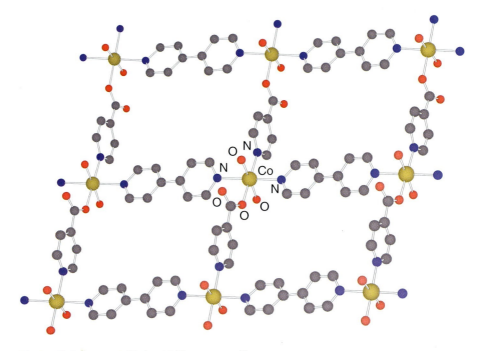

21

self-assembly of the respective components (metal nitrate, 4,4'-bipyridine and pyridine-4-carboxylic acid) in aqueous solution to yield a species of empirical formula $[M(4,4'\text{-bipyridine})(\text{pyridine-4-carboxylate})(H_2O)_2]^+$ [M = Co(II) or Cd(II)] which crystallises in the grid-like structure illustrated in the figure. The design of this framework relies on two unique edge lengths being employed; this was achieved by using a combination of neutral and charged bridging ligands in the structure. The individual grids form stacked layers in the crystal in which hydrogen bonds and π–π interactions align the layers to produce interconnected microchannels that run approximately perpendicular to the layers. These channels are occupied by disordered nitrate ions and disordered water molecules that are hydrogen-bound

Figure 7.4 *An assembled grid-like structure*[33]

together. The preparation and properties of other types of supramolecular grids are discussed in a later section of this chapter.

7.2.3 Related Polygonal and Polyhedral Systems

Application of the lessons learnt in the self-assembly of molecular squares has led to successful strategies for preparing further molecular polygons (and, in particular, new triangles, rectangles and hexagons) as well as examples of particular other polyhedra.[34–37]

To construct a hexagon, the shape-defining corner units need to present 120° angles to the linker groups. With this in mind, a corner unit derived from **22** (Figure 7.5) was employed in one study by Stang *et al.*[6] Addition of equimolar amounts of 4,4'-bipyridine to **22** in dichloromethane, followed by stirring for half an hour at room temperature, resulted in a quantitative yield of the required hexagon **23**. The feasibility of using a range of other linker molecules in such a reaction was also demonstrated in this study.

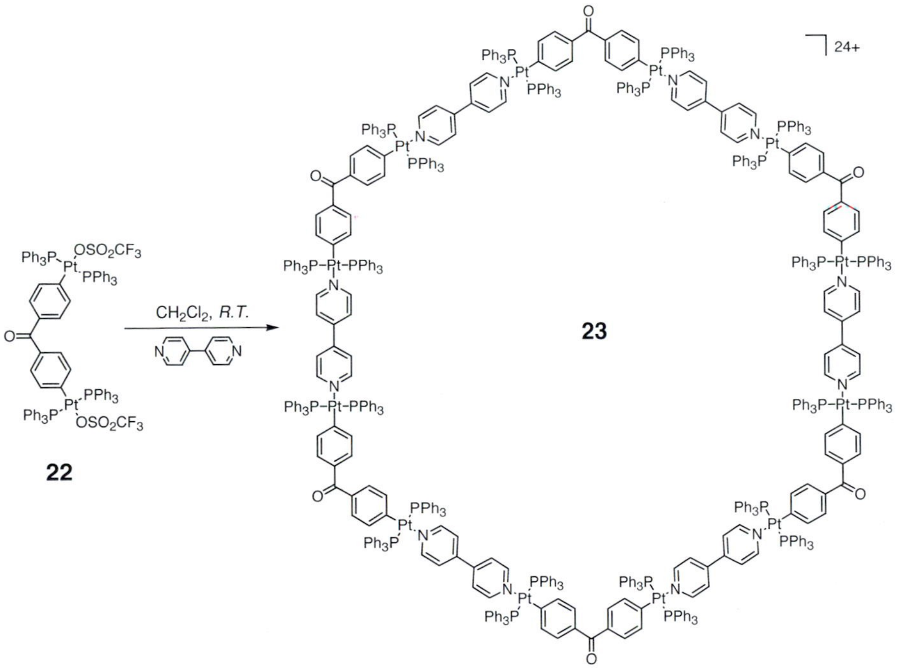

Figure 7.5 *Self-assembly of the hexagon* **23**[6]

34 D.H. Cao, K. Chen, J. Fan, J. Manna, B. Olenyuk, J.A. Whiteford and P.J. Stang, *Pure Appl. Chem.*, 1997, **69**, 1979.
35 P.J. Stang and B. Olenyuk, *Acc. Chem. Res.*, 1997, **30**, 502.
36 B. Olenyuk, A. Fechtenkotter and P.J. Stang, *J. Chem. Soc., Dalton Trans.*, 1998, 1707.
37 P.J. Stang, *Chem. Eur. J.*, 1998, **4**, 19.

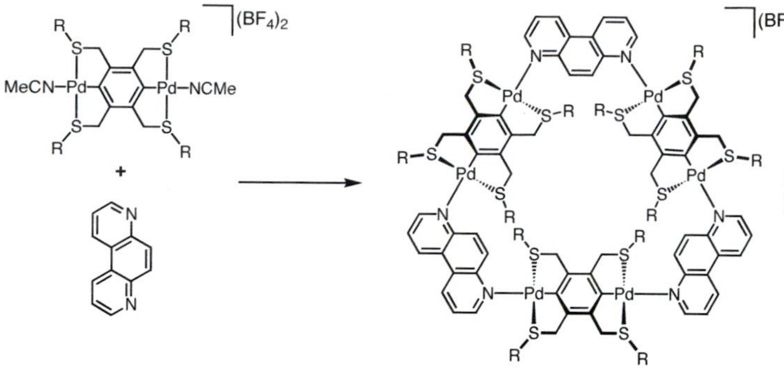

Figure 7.6 *Assembly of the molecular hexagons* **24a** *and* **24b**[38]

A further type of hexagonal complex, represented by **24a** and **24b** in Figure 7.6 has been generated using the somewhat different synthetic approach of employing a linear dinuclear metal component linked by 4,7-phenanthroline.[38] The latter is a commercially available ligand that clearly promotes bridging instead of simple chelation; it provides a rigid 60° corner suitable for construction of hexagonal arrays. The metal-containing walls in this structure are orientated approximately at right angles to the three corner pieces, with all six palladium atoms being constrained to lie in a plane. Pd–Pd–Pd angles are near 120° in each case.

The tendency to build upon knowledge gained in previous studies to produce ever more elaborate structures is well illustrated by the self-assembly process summarised in Figure 7.7.[39] Using this, the new nanometre-sized derivative of palladium(II), **25**, was obtained. In this case the product forms spontaneously by the assembly of a total of nine subunits. The macrotricyclic nature of **25** has been confirmed by X-ray diffraction.

The generation of a polyhedral supramolecular species is further exemplified by the spontaneous self-assembly of a tetrahedral host–guest complex from the reaction of six fumaronitrile (NCC_2H_2CN) ligands, four iron(II) ions, four $CH_3C(CH_2PPh_2)_3$ tripod ligands and a tetrafluoroborate anion.[40] The structure of the cationic product (Figure 7.8) indicates that it exhibits C_2 symmetry; the symmetry of the system is reflected by the positions of the four iron centres at the corners of the tetrahedron. The B–F bonds of the encapsulated BF_4^- point towards the four iron atoms which adopt their usual octahedral co-ordination. From the close match of host and anionic guest, it seems likely that the latter plays a templating role in the assembly of the former.

The self-assembly of a molecular square consisting of four [Pt(ethane-1,2-diamine)]$^{2+}$ corner units and four singly deprotonated uracil nucleobases **26** has

[38]   J.R. Hall, S.J. Loeb, G.K.H. Shimizu and G.P.A. Yap, *Angew. Chem., Int. Ed.*, 1998, **37**, 121.
[39]   M. Fujita, S.-Y. Yu, T. Kusukawa, H. Funaki, K. Ogura and K. Yamaguchi, *Angew. Chem., Int. Ed. Engl.*, 1998, **37**, 2082.
[40]   S. Mann, G. Huttner, L. Zsolnai and K. Heinze, *Angew. Chem., Int. Ed. Engl.*, 1996, **35**, 2808.

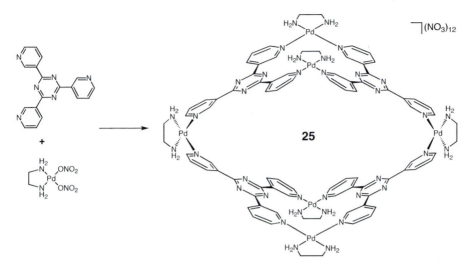

Figure 7.7 *Assembly of the macrotricyclic metallocycle* **25**[39]

been reported.[41,42] This unusual species, namely [Pt(ethane-1,2-diamine)(UH$_{-1}$)-(H$_2$O)]$^+$, forms spontaneously from the related monomeric platinum(II) derivative of ethane-1,2-diamine and singly deprotonated uracil (UH$_{-1}$). This tetranuclear cationic product shows close similarity to the calix[4]arenes as far as overall geometry and conformational behaviour is concerned. In the solid state, a 1,3-alternate (up–down) arrangement of the four uracil rings is present, whereas in solution a second major configuration – a cone arrangement – is generated. The new tetranuclear product appears to show an enhanced affinity for binding additional metal cations. Thus, this product undergoes further deprotonation coupled with uptake of four additional transition metal species to yield new octa-metal squares of type [{Pt(ethane-1,2-diamine)X(UH$_{-2}$)}$_4$]$^{8+}$ [where X = *cis*-Pt^{II}(NH$_3$)$_2$; Pt^{II}(ethane-1,2-diamine); Ni^{II}(H$_2$O)$_2$; Pd^{II}(ethane-1,2-diamine), and Cu^{II}].[43] A related cationic product of type [Pt(ethane-1,2-diamine)Ag(UH$_{-1}$)$_4$]$^{8+}$ was also prepared. These products have a second metal ion located at each corner of their structures; the X-ray structure for the product with X = *cis*-Pt^{II}(NH$_3$)$_2$ is illustrated in Figure 7.9. In this case, there is a 1,3-*alternate* arrangement of the four uracil anions, each with its four heteroatoms bound to metal centres.

26, Uracil (U)

41 H. Rauter, E.C. Hillgeris, A. Erxleben and B. Lippert, *J. Am. Chem. Soc.*, 1994, **116**, 616.
42 H. Rauter, E.C. Hillgeris and B. Lippert, *J. Chem. Soc., Chem. Commun.*, 1992, 1385.
43 H. Rauter, I. Mutikainen, M. Blomberg, C.J.L. Lock, P. Amo-Ochoa, E. Freisinger, L. Randaccio, E. Zangrando, E. Chairparin and B. Lippert, *Angew. Chem., Int. Ed. Engl.*, 1997, **36**, 1296.

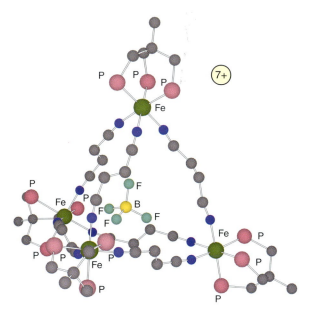

Figure 7.8 *X-Ray structure of a tetrahedral metallocycle including a tetrafluoroborate guest[40]*

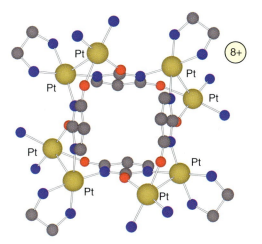

Figure 7.9 *X-Ray structure of the {[Pt(ethane-1,2-diamine)(cis-Pt(NH₃)₂)(uracil-2H)]₄}⁸⁺ cation[43]*

The self-assembly of the supramolecular cube **27** from 20 components has been reported.[44] Using nitromethane as a non-co-ordinating solvent, [Ru([9]ane-S₃)(dimethyl sulfoxide)Cl₂] was treated with 4,4-bipyridine in an 8 : 12 (namely 2 : 3) ratio and the reaction was monitored by NMR spectroscopy. The proposed

⁴⁴   S. Roche, C. Haslam, H. Adams, S.L. Heath and J.A. Thomas, *Chem. Commun.*, 1998, 1681.

product **27** was observed to form over four weeks. The formation of this symmetrical cube was anticipated on the basis, discussed previously, that both enthalpic and entropic arguments suggest that discrete supramolecular architectures will be favoured over polymeric products.

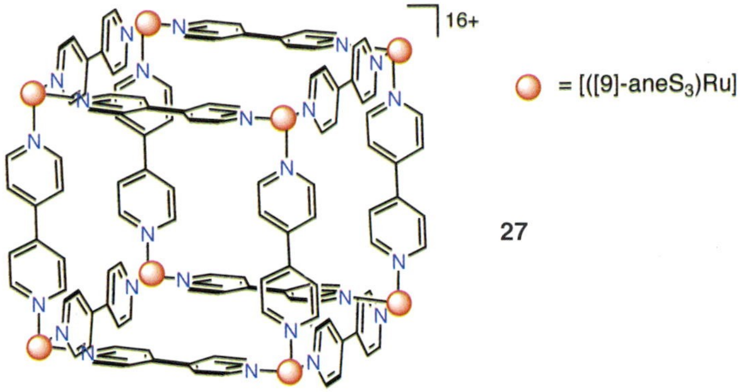

$\bigcirc$ = [([9]-aneS$_3$)Ru]

27

7.2.4 A Further Cage-like Structure

A template strategy has been employed to produce a three-dimensional cage-like assembly formed from the tripyridyl derivative **28** and dinitro(ethane-1,2-diamine) palladium(II).[45] The interaction of these species initially led to an intractable mixture of oligomeric products. However, addition of a suitable guest, such as (*p*-methoxyphenyl)acetate, to the reaction mixture resulted in the NMR signals for the oligomers gradually disappearing and new signals corresponding to the desired host–guest complex of the cage **29** appearing. The assembly of the latter was complete within several hours, with the product being isolated in 94% yield. Clearly, the guest serves as a template for the assembly of this unusual multi-component complex.

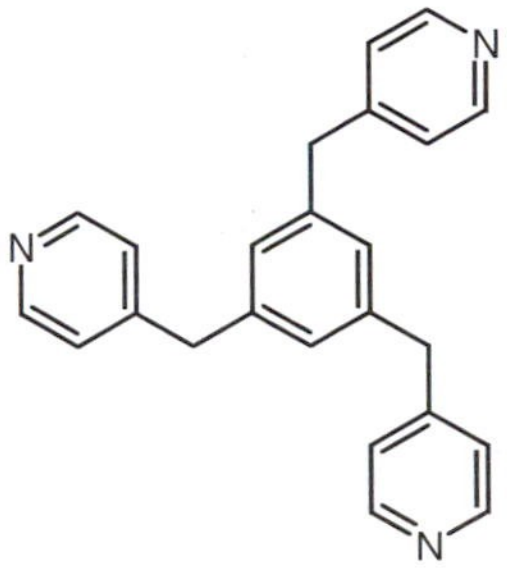

28

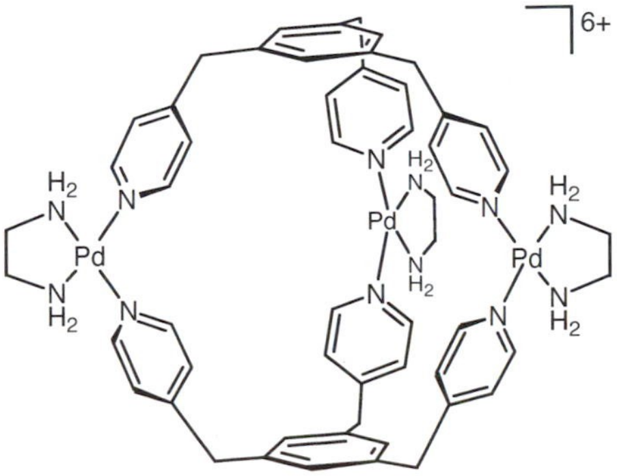

29

[45]   M. Fujita, S. Nagao and K. Ogura, *J. Am. Chem. Soc.*, 1995, **117**, 1649.

7.2.5   A Linked Metal-cluster System

The use of 'programmed' ligands to produce designed molecular architectures *via* conventional metal–ligand interactions has, of course, been a feature of much of the work described so far in both this and the previous chapter. Metal-cluster formation is another area which, although differing subtlely from the above, often also involves the programmed assembly of multi-component polynuclear species. As mentioned in Chapter 1, no attempt will be made to survey the vast chemistry of metal-cluster compounds here. However, a system crossing both areas has been reported by Hendrickson, Christou *et al.*[46] These workers employed the bis-2,2'-bipyridine ligand **30** (Figure 7.10) for 'carboxylato-manganese' cluster formation. Thus, treatment of $[Mn_3O(O_2CEt)_6(pyridine)_3](ClO_4)$ with **30** in a 1 : 1.5 molar ratio in dichloromethane led to the subsequent isolation of the octanuclear complex $[Mn_8O_4(O_2CEt)_{14}(L)_2](ClO_4)_2$ (L = **30**) whose X-ray structure is shown in Figure 7.10. Each bis-bipyridine ligand links two metal-cluster centres, composed of butterfly-like $[Mn_4O_2(O_2CEt)_7]^+$ cores, to give a large, cyclic, centrosymmetric cation incorporating a central void. The formation of this latter species can thus be seen to be controlled by two 'programmes' – one influencing the formation of the

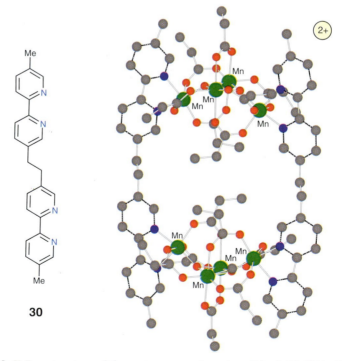

30

Figure 7.10 *X-Ray structure of the centrosymmetric cation, $[Mn_8O_4(O_2CEt)_{14}(L)_2]^{2+}$ (L = **30**)[46]*

[46]   V.A. Grillo, M.J. Knapp, J.C. Bollinger, D.N. Hendrickson and G. Christou, *Angew. Chem., Int. Ed. Engl.*, 1996, **35**, 1818.

$[Mn_4(\mu_3\text{-}O)_2]^{8+}$ centres, while the second dictates the assembly of the 'dimer-of-clusters' arrangement.

7.2.6 Catenated Metallocycles

Catenated systems incorporating metal-containing rings are rare.[47] Nevertheless, systems of this type involving organometallic bond formation are known.[48–51] In one such example, **31** assembles from its two preformed rings *via* a sequence that involves Mg–C bond dissociation followed by threading of the crown ether ring by the corresponding open-chain magnesium derivative; reformation of a metal–carbon bond then occurs.[48,49]

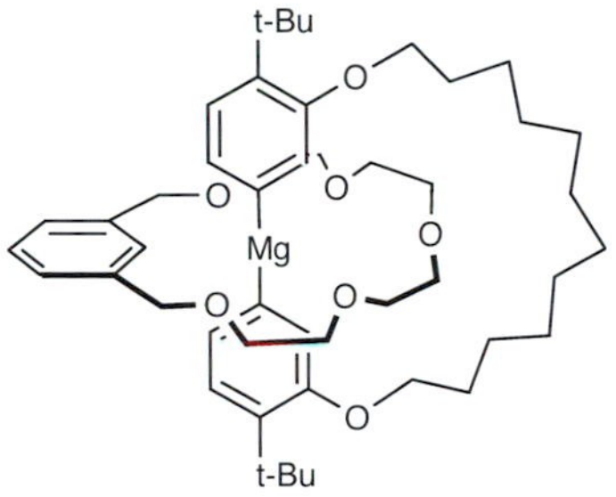

31

Semi-flexible bipyridine derivatives incorporating alkyl spacer groups between the pyridyl rings have been demonstrated to yield dinuclear palladium(II) species related to those discussed in Section 7.2 of this chapter.[16,52] Thus, insertion of $-CH_2CH_2-$ or $-CH_2(C_6F_4)CH_2-$ groups between the 4- and 4'-positions in the parent bipyridine ligand leads to the formation of 2 : 2 (metal : ligand) complexes of type **32** on reaction of these 'extended' bipyridine derivatives with dinitro(ethane-

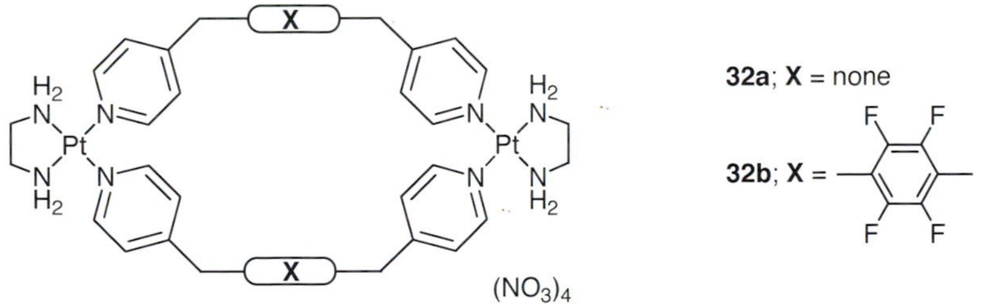

32a; X = none

32b; X =

47   A. Grohmann, *Angew. Chem., Int. Ed. Engl.*, 1995, **34**, 2107.
48   G.-J.M. Gruter, F.J.J. de Kanter, P.R. Markies, T. Nomoto, O.S. Akkerman and F. Bickelhaupt, *J. Am. Chem. Soc.*, 1993, **115**, 12179.
49   P.R. Markies, O.S. Akkerman, F. Bickelhaupt, W.J.J. Smeets and A.L. Spek, *Organometallics*, 1994, **13**, 2616.
50   D.M.P. Mingos, J. Yau, S. Menzer and D.J. Williams, *Angew. Chem., Int. Ed. Engl.*, 1995, **34**, 1894.
51   D.M.P. Mingos, *J. Chem. Soc., Dalton Trans.*, 1996, 561.
52   M. Fujita, S. Nagao, M. Iida, K. Ogata and K. Ogura, *J. Am. Chem. Soc.*, 1993, **115**, 1574.

1,2-diamine)palladium(II). The increased flexibility of the bipyridine derivatives apparently allows the pyridyl lone pairs to attain a suitable orientation for generation of a dinuclear ellipsoid geometry – the latter being favoured over a tetra-nucleated square arrangement. Both compounds of type **32** also show an ability for molecular recognition of electron-rich aromatic compounds, with the tetrafluorophenylene-containing derivative yielding especially strong adducts (as might be expected from the increased electron-deficient nature of its cavity due to the influence of the electron-withdrawing tetrafluorophenylene groups).

An interesting feature of the dinuclear palladium derivative **33** (Figure 7.11) is that it exists in solution in rapid equilibrium with the corresponding catenane **34**.[53,54] The formation of the latter is favoured in solution as the concentration is increased. This behaviour appears to reflect the presence of 'double molecular recognition' in which two molecules of **33** both recognise and bind each other in their respective cavities. Based on NMR evidence, a mechanism involving initial association of two rings of **33** followed by two sequential ligand exchanges between these molecular rings, concomitant with a twisting of the ligand strands around each other, has been proposed.[55]

A remarkable medium effect was also demonstrated to influence the position of the above equilibrium.[53] Employment of a more polar medium (a D_2O solution of sodium nitrate) pushed the equilibrium strongly towards the catenated form, with greater than 99% of this form being present, even at relatively low concentrations of the above salt. The observation that the change in the equilibrium position is rapid confirmed the inherent lability of the palladium(II)–pyridine linkages in this

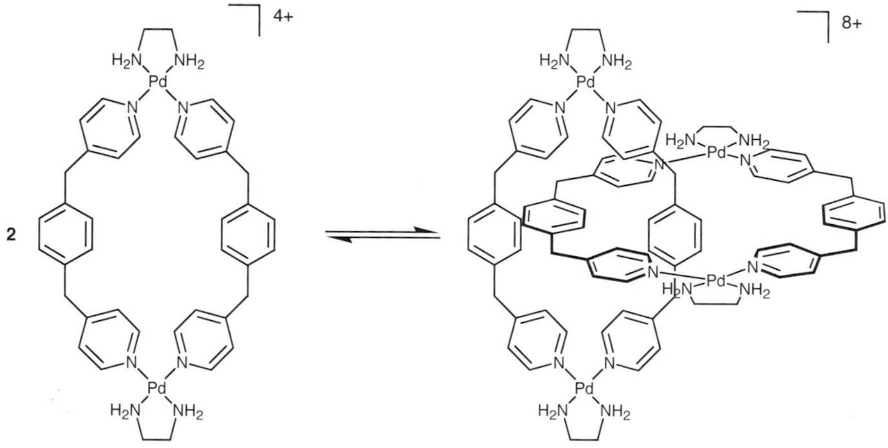

Figure 7.11 *Formation of the palladium-containing [2]-catenane* **34**[59]

53 M. Fujita, F. Ibukuro, H. Hagihara and K. Ogura, *Nature*, 1994, **367**, 720.
54 M. Fujita, F. Ibukuro, K. Yamaguchi and K. Ogura, *J. Am. Chem. Soc.*, 1995, **117**, 4175.
55 M. Fujita, F. Ibukuro, H. Seki, O. Kamo, M. Imanari and K. Ogura, *J. Am. Chem. Soc.*, 1996, **118**, 899.

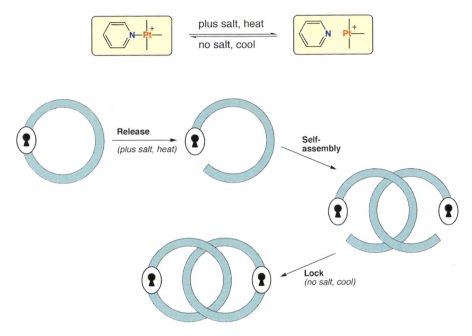

Figure 7.12 *Controlling the 'molecular lock'*[53]

system. It was realised by Fujita *et al.* that if the metal–nitrogen bond could be frozen after the catenane had assembled, then this would result in a stable linked system. Such an outcome was achieved with the corresponding platinum(II) system by employing the concept of a 'molecular lock' (Figure 7.12).[54] The platinum(II)–pyridine bond can be considered to be a lock since it is non-labile (locked) under normal conditions, but becomes labile (unlocked) in highly polar media (aqueous $NaNO_3$) and at elevated temperatures. Under high temperature conditions, the catenane is free to self-assemble; it may then be 'locked' in position by simply removing the salt with simultaneous cooling of the solution.

7.2.7 Porphyrin-containing Metallocycles

The rich chemistry of metal porphyrins, as well as their general robustness to chemical degradation, has made the incorporation of these macrocyclic systems into ordered molecular assemblies especially attractive. Further, the photochemical properties of porphyrins impart a functionality to the resulting assemblies that may, for example, be used to aid their characterisation.

Following earlier work on the self-assembly of a porphyrin cage (involving molecular recognition through hydrogen-bond formation)[56] as well as the use of pyridyl-derivatised porphyrins as subunits for the formation of polymeric net-

[56]   C.M. Drain, R. Fischer, E.G. Nolen and J.-M. Lehn, *J. Chem. Soc., Chem. Commun.*, 1993, 243.

35a; M = 2H
35b; M = Zn(II)

36a; M = 2H
36b; M = Zn(II)

works,[57] Drain and Lehn[58] synthesised the impressive new tetraporphyrin square arrays **35** and **36**. The success of these syntheses, which involve self-assembly *via* reaction of *cis* or *trans* meso-dipyridyl porphyrin derivatives with *cis-* or *trans-* $M(NCPh)_2Cl_2$ (with loss of the NCPh ligands), demonstrated that related strategies to those employed in the earlier studies are applicable to the formation of assemblies having substantially higher molecular weights.

Other large supramolecular systems incorporating (circularly) linked porphyrin rings have been reported. An example of this type, **37** (see Figure 7.13), was reported by Sanders *et al.*[3p,59] The preparation of this novel species proceeds *via* the initial formation of a linked zinc porphyrin dimer, which then condenses with a further zinc porphyrin entity in the presence of tris(pyridylacetoacetato)aluminum(III). The latter acts as a triangular template for arranging the above zinc porphyrin derivatives close enough for coupling to occur so that a cyclic product containing three alkyne–platinum–alkyne linkages results. This product has the extraordinary empirical formula $C_{210}H_{246}P_6O_6Pt_3Zn_3Al$. Removal of the zinc(II) ions from each porphyrin leads to the release of the cyclic trimer from its template.

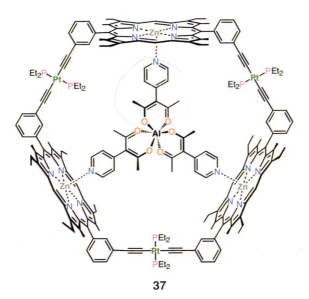

37

Figure 7.13 *Zinc porphyrin-containing assembled array formed using tris(pyridylaceto-acetato)aluminium(III) as a template*[59]

57 B.F. Abrahams, B.F. Hoskins and R. Robson, *J. Am. Chem. Soc.*, 1991, **113**, 3606; E.B. Fleischer and A.M. Shachter, *Inorg. Chem.*, 1991, **30**, 3763; Y. Kobuke and H. Miyaji, *J. Am. Chem. Soc.*, 1994, **116**, 4111.

58 C.M. Drain and J.-M. Lehn, *J. Chem. Soc., Chem. Commun.*, 1994, 2313.

59 (a) H.L. Anderson and J.K.M. Sanders, *Angew. Chem., Int. Ed. Engl.*, 1990, **29**, 1400; (b) L.G. Mackay, H.L. Anderson and J.K.M. Sanders, *J. Chem. Soc., Perkin Trans. 1*, 1995, 2269.

7.2.8 Higher Oligonuclear Metallocycles

Another group of self-assembled metallocycles is represented by the cyclic dode-canuclear complex **38** (Figure 7.14). This species was obtained by reaction of hydrated nickel(II) acetate with molten 6-chloro-2-pyridone followed by extraction of the reaction product with tetrahydrofuran.[60] This interesting metallocycle was isolated as green crystals. Further cyclic examples of this general type include a series of octanuclear copper complexes incorporating the same mixture of ligands,[61] a related dodecanuclear cobalt complex that also contains mixed acetate and pyridone bridges[62] and the decanuclear 'ferric wheel', $[Fe(OMe)_2(O_2CCH_2Cl)]_{10}$, prepared by Lippard *et al.*[63]

Reaction of 'cobalt(III) acetate' and ammonium hexafluorophosphate in methanol over one week at room temperature yielded a mixture of products that included green crystals of $[NH_4][Co_8(MeCO_2)_8(OMe)_{16}][PF_6]$.[64] The cavity in this product

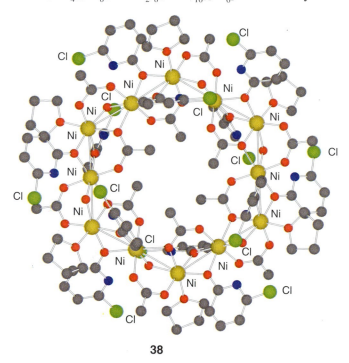

38

Figure 7.14 *X-Ray structure of the dodecanuclear complex* **38**[60]

[60] A.J. Blake, C.M. Grant, S. Parsons, J.M. Rawson and R.E.P. Winpenny, *J. Chem. Soc., Chem. Commun.*, 1994, 2363.
[61] A.J. Blake, C.M. Grant, P.E.Y. Milne, J.M. Rawson and R.E.P. Winpenny, *J. Chem. Soc., Chem. Commun.*, 1994, 169.
[62] W. Clegg, C.D. Garner and M.H. Al-Samman, *Inorg. Chem.*, 1983, **22**, 1534.
[63] K.L. Taft, C.D. Delfs, G.C. Papaefthymiou, S. Foner, D. Gatteshi and S.J. Lippard, *J. Am. Chem. Soc.*, 1994, **116**, 823.
[64] J.K. Beattie, T.W. Hambley, J.A. Klepetko, A.F. Masters and P. Turner, *J. Chem. Soc., Chem. Commun.*, 1988, 45.

functions as a host for the ammonium cation, with charge balance being provided by the hexafluorophosphate counter-ion. This highly symmetric toroidal species has its eight cobalt atoms bridged by methoxy and acetato ligands to form a planar ring.

Other cyclic structures related to that of $[Co_8(MeCO_2)_8(OMe)_{16}]$ have been reported: $[V_8O_8(OMe)_{16}(C_2O_4)]^{2+}$ and $[V_{12}O_{32}]^{4-}$,[65,66] $[Cr_8F_8(Me_3CCO_2)_{16}]$,[67] $[Mo_8O_{16}(OR)_8(C_2O_4)]^{2+}$ (R = Me, Et),[65,68] $[Fe_8F_8(Me_3CCO_2)_{16}]$,[69] $[Cu_8L_8(OH)_8]$ (L = 3,5-dimethylpyrazole)[70] and $[Fe_6Cl_6\{CH_3N(CH_2CH_3O)_2\}]$.[71] In some of these systems, solvent molecules were found to occupy the central cavities.

Although the presence of mixed bridging ligands may favour the formation of high nuclearity assemblies, it remains difficult to pinpoint the factors that give rise to these unusual cyclic products. In accordance with the previous discussion, it seems likely that these multi-component species may exist in solution in equilibrium with related oligomeric species; their isolation will thus be influenced by both the position of the equilibrium as well as by the relative solubilities of the respective species in the chosen reaction solvent.

A feature of several of the cyclic derivatives of the present type incorporating transition metal ions is that they display communication between their metal centres. For example, initial magnetic studies involving **38** suggest the presence of ferromagnetic coupling between the metal centres to yield an $S = 12$ ground state; this was claimed to represent one of the highest spin ground states yet observed for a molecular species.[60]

7.3   Co-ordination Arrays

The metal template assembly of co-ordination arrays corresponding to from one- to three-dimensions have now all been well demonstrated.

7.3.1   One-dimensional Systems

There are now numerous, metal-linked oligomeric (and polymeric) systems that fall into this category. For example, the acetylacetonates of manganese(II), nickel(II) and zinc(II) have long been known to be trimeric while the cobalt(II) complex is tetrameric, with three β-diketonate oxygen atoms bridging adjacent metal centres in a linear array in each case.[72] Other more recent examples include systems built

65 Q. Chen, S. Liu and J. Zubieta, *Inorg. Chem.*, 1989, **28**, 4433.
66 V.W. Day, W.G. Klemperer and O.M. Yaghi, *J. Am. Chem. Soc.*, 1989, **111**, 5959.
67 N.V. Gerbeleu, Y.T. Struchkov, G.A. Timko, A.S. Batsanov, K.M. Indrichan and D.A. Popovich, *Dokl. Acad. Nauk SSSR*, 1990, **313**, 1459.
68 Q. Chen, S. Liu and J. Zubieta, *Angew. Chem., Int. Ed. Engl.*, 1988, **27**, 1724.
69 N.V. Gerbeleu, Y.T. Struchkov, O.S. Manole, G.A. Timko and A.S. Batsanov, *Dokl. Akad. Nauk SSSR*, 1993, **331**, 184.
70 G.A. Ardizzoia, M.A. Angaroni, G.L. Monica, F. Cariati, M. Moret and N. Masciocchi, *J. Chem. Soc., Chem. Commun.*, 1990, 1021.
71 R.W. Saalfrank, I. Bernt, E. Uller and F. Hampel, *Angew. Chem., Int. Ed. Engl.*, 1997, **36**, 2482.
72 F.A. Cotton and G. Wilkinson, *Advanced Inorganic Chemistry*, John Wiley & Sons, New York, 5th edn., 1988, p. 478.

39; M = Ru(II), *n* = 2
40; M = Rh(III), *n* = 6

around, for example, two terpyridine fragments linked back-to-back by a rigid or semi-rigid spacer – which may or may not contain additional functionality (such as a porphyrin derivative).[73–75] Such systems, of which **39** and **40** are examples of the latter type,[76,77] are terminated at each end by co-ordination to an octahedral metal such as ruthenium(II), osmium(II) or rhodium(III). The co-ordination sphere of each metal is then typically completed (as in the above example) by attachment to a 'monomeric' terpyridine moiety. A motivation for the design and synthesis of a wide range of species of this type [including systems based on rigidly linked bis-bipyridyl units,[78] bis-2-(2′-pyridyl)benzimidazolyl units,[79] bridging 2,3,5,6-tetra(2-pyridyl)pyrazine units[80] and metal-linked porphyrin derivatives[81]] has often centred around the production of supramolecules that are capable of forming charge-separated species following photonic excitation,[82] or which might act as molecular 'wires'.[83] Alternatively, the inclusion of different metals [for example, ruthenium(II) and osmium(II)] in each of the terminal metal sites enables strict control of their internuclear distance and hence provides a means of influencing any electrochemically-induced, energy transfer process occurring between them.

The tris-terpyridine system **41** (Figure 7.15), incorporating –$(CH_2)_{10}$– spacers

[73] E.C. Constable, A.M.W. Cargill Thompson and D.A. Tocher, in *Supramolecular Chemistry*, eds. V. Balzani and L. De Cola, Kluwer Academic Publishers, Dordrecht, 1992, p. 219.
[74] E.C. Constable and A.M.W. Cargill Thompson, *J. Chem. Soc., Dalton Trans.*, 1992, 3467; F. Barigelletti, L. Flamigni, V. Balzani, J.-P. Collin, J.-P. Sauvage, A. Sour, E.C. Constable and A.M.W. Cargill Thompson, *J. Chem. Soc., Chem. Commun.*, 1993, 942; E.C. Constable, A.M.W. Cargill Thompson and S. Greulich, *J. Chem. Soc., Chem. Commun.*, 1993, 1444; J.-P. Collin, P. Laine, J.-P. Launay, J.-P. Sauvage and A. Sour, *J. Chem. Soc., Chem. Commun.*, 1993, 434; F. Barigelletti, L. Flamigni, V. Balzani, J.-P. Collin, J.-P. Sauvage, A. Sour, E.C. Constable, A.M.W. Cargill Thompson, *J. Am. Chem. Soc.*, 1994, **116**, 7692.
[75] F. Barigelletti, L. Flamigni, V. Balzani, J.-P. Collin, J.-P. Sauvage, A. Sour, E.C. Constable and A.M.W. Cargill Thompson, *Coord. Chem. Rev.*, 1994, **132**, 209; E.C. Constable, C.E. Housecroft and L.A. Johnston, *Inorg. Chem. Commun.*, 1998, **1**, 68.
[76] J.-P. Collin, A. Harriman, V. Heitz, F. Odobel and J.-P. Sauvage, *J. Am. Chem. Soc.*, 1994, **116**, 5679.
[77] J.-P. Collin, P. Gavina, V. Heitz and J.-P. Sauvage, *Eur. J. Inorg. Chem.*, 1998, 1 and references therein.
[78] P. Belser, *Chimia*, 1994, **48**, 347; A.I. Baba, J.R. Shaw, J.A. Simon, R.P. Thummel and R.H. Schmehl, *Coord. Chem. Rev.*, 1998, **171**, 43.
[79] K. Nozaki and T. Ohno, *Coord. Chem. Rev.*, 1994, **132**, 215.
[80] E.C. Constable, A.J. Edwards, D. Phillips and P.R. Raithby, *Supramolecular Chem.*, 1995, **5**, 93.
[81] E. Alessio, M. Macchi, S.L Heath and L.G. Marzilli, *Inorg. Chem.*, 1997, **36**, 5614 and references therein.
[82] V. Balzani, A. Credi and M. Venturi, *Coord. Chem. Rev.*, 1998, **171**, 3.
[83] A. Harriman and R. Ziessel, *Coord. Chem. Rev.*, 1998, **171**, 331.

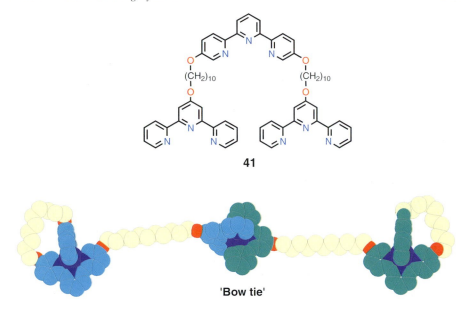

41

'Bow tie'

Figure 7.15 *X-Ray structure of the 'bow tie' system $[Fe_3L]^{6+}$ (L = **41**)*[84]

between the terpyridine units, reacts with iron(II) chloride in methanol–chloroform to yield a species of type $[Fe_3L_2]^{6+}$.[84] Owing to the presence of the flexible spacers, a helical geometry is not promoted in this case. Similarly, while the flexibility of the ligand system would appear to allow the formation of a [2]-catenane in which two metals are used to generate interlocked metallocycles (that are held in place by the binding of the third metal to the centre terpyridyl of each ligand), this also does not occur. Instead, this complex cation adopts the alternative double-looped ('bow tie') arrangement illustrated in (Figure 7.15).

7.3.2 Two-dimensional Systems

Finite, two-dimensional systems may be broadly divided into three categories. In terms of increasing complexity these are: molecular racks, ladders and grids (Figure 7.16).

The synthesis of the simple dinuclear racks **42–45**, incorporating two bis-terpyridine ruthenium(II) co-ordination spheres, has been achieved by reaction of 2.5 equivalents of $[Ru(2,2':6',2'$-terpyridine)Cl_3]$ with the appropriate bis-tridentate species, followed by addition of hexafluorophosphate anion.[85] X-Ray diffraction studies have allowed elucidation of the effect of the central R group on the overall shape of the respective molecular racks. Thus, when R = Me (**43**) there is a straightening of the structure relative to the presence of more bulky alkyl substituents. It

[84] E.C. Constable and D. Phillips, *Chem. Commun.*, 1997, 827
[85] G.S. Hanan, C.R. Arana, J.-M. Lehn, G. Baum and D. Fenske, *Chem. Eur. J.*, 1996, **2**, 1292 and references therein.

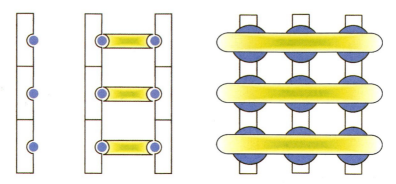

Figure 7.16 *A molecular rack, ladder and grid*

was suggested that a smaller substituent (the derivative **42** with R = H was subsequently synthesised – see below) should yield a molecular rack in which the ancillary ligands are almost parallel. Structures such as these (especially when extended racks are involved) are of interest for potential use in molecular electronics since they provide the prospect of redox-moderated communication occurring between metal centres along the rack.

42; R = H
43; R = CH₃
44; R = —phenyl
45; R = —anthryl

46

X = anthryl

In an extension of this study, the analogous complexes with R = H (**42**) and 9-anthryl (**44**) were prepared. The crystal structure of derivative **44** has been determined. Once again, the generation of an ordered, rack-type structure is revealed with steric interactions between the anthryl moieties and the ancillary terpyridyl ligands in evidence.[86] A related self-assembly procedure employing the corresponding tris-tridentate derivative led to **46** being obtained in 58% yield.

[86]   G.S. Hanan, C.R. Arana, J.-M. Lehn and D. Fenske, *Angew. Chem., Int. Ed. Engl.*, 1995, **34**, 1122.

Following directly from the initial part of the above study, Lehn *et al.*[87] employed sequential complexation strategies to introduce different octahedral metal ions into 2 × 2 molecular grids of the type shown in Figure 7.17; each of these incorporates four molecules of **47**. An important feature of the procedure was to introduce the respective metal ions sequentially in order of their increasing lability; this was necessary to avoid scrambling of the metal centres. For this reason, ruthenium(II) or osmium(II) were introduced before iron(II), cobalt(II) or nickel(II).

For the preparation of the required 1 : 2 [Ru : ligand **47**] intermediate, a protection–deprotection strategy was employed. This involved reaction of **47** with trimethyloxonium tetrafluoroborate to block one metal co-ordination site in **47** to yield **48**. After co-ordination of one ruthenium ion, the remaining site was unblocked by reaction with DABCO in refluxing acetonitrile. Reaction of this 1 : 2 (metal : ligand) product with a second metal ion was then carried out.

In the case of the syntheses of the osmium-containing species, reaction of ammonium hexachloro-osmate with two equivalents of **47** in refluxing ethylene glycol proceeded smoothly to yield the required 1 : 2 (Os : ligand) intermediate directly in 40% yield. The higher kinetic inertness of osmium(II) complexes relative to those of ruthenium(II) may be the reason that the complexation could be stopped after the uptake of one metal ion in this synthesis. Once again, reaction of this intermediate with further metal ion(s) led to formation of the required mixed metal species (Figure 7.17).

In a similar vein, self-assembly of the dimethyl derivative of the corresponding tris(bipyridine) ligand in the presence of bipyrimidine and copper(I) spontaneously generates the new 'ladder' structure **49**.[88] A similar tetranuclear structure derived from the corresponding bis(bipyridine) ligand has also been reported.

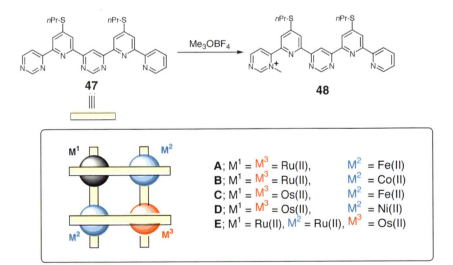

Figure 7.17 *Mixed-metal molecular grids*[87]

[87] D.M. Bassani, J.-M. Lehn, K. Fromm and D. Fenske, *Angew. Chem., Int. Ed. Engl.*, 1998, **37**, 2364.
[88] P.N.W. Baxter, G.S. Hanan and J.-M. Lehn, *Chem. Commun.*, 1996, 2019.

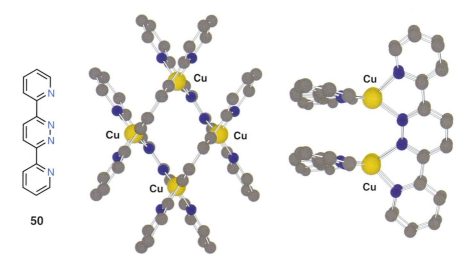

49

Systems exhibiting polynuclear grid-like architectures, constructed from di- or tri-topic ligands incorporating 'side-by-side' metal co-ordination sites, have been reported.[89–91] For example, the structure shown in Figure 7.18 forms on interaction of the difunctional ligand **50** with copper(I).[89] This semi-rigid structure contains a planar arrangement of its four copper atoms; these lie on the vertices of a slightly

Figure 7.18 *A polynuclear grid architecture incorporating the difunctional ligand* **50**[90]

[89]   M.-T. Youinou, N. Rahmouni, J. Fischer and J.A. Osborn, *Angew. Chem., Int. Ed. Engl.*, 1992, **31**, 733.
[90]   P.N.W. Baxter, J.-M. Lehn, B.O. Kneisel and D. Fenske, *Chem. Commun.*, 1997, 2231.
[91]   G.S. Hannan, D. Volkmer, U.S. Schubert, J.-M. Lehn, G. Baum and D. Fenske, *Angew. Chem., Int. Ed. Engl.*, 1997, **36**, 1842.

distorted rhombus. The copper(I) atoms are spanned by four ligands, with two lying above the Cu_4 mean plane and two below. Each copper(I) ion has a distorted tetrahedral environment arising from binding to two ligands, each of which contribute a pyridine and a pyridazine nitrogen donor. The ligands of each set adopt a parallel orientation in which the separation is 3.47 Å, apparently reflecting significant π–π stacking between adjacent aromatic rings.

Two-dimensional, metal-containing grids are of particular interest as their structures suggest the basis for the construction of information storage devices. Arrays on this type, in which the metal ions act as ion dots that might prove accessible by external stimuli, would be of smaller size than quantum dots. They have the potential advantage of self-assembling spontaneously and hence may eliminate the need for microfabrication.[91]

The 2×2 grid **52** formed from four molecules of **51** and four copper(I) ions has been reported to form in almost quantitative yield.[90] The internal box-like cavity is a rhombus rather than a square, reflecting deviation from perfect tetrahedral coordination about each copper(I) ion. The crystal structure revealed that six guests, four benzenes and two nitromethanes, were trapped in the grooves defined by the pairs of parallel ligands. The presence of the terminal phenyl rings on the ligands appears necessary for grid formation since in their absence 1 : 1 species form instead.

51   4 Cu(I)   **52**

A related 3×3 grid consisting of an array of nine silver(I) ions and six trifunctional ligands of type **53** has been reported.[92] The ^{109}Ag NMR spectrum of this product (not illustrated) proved to be quite useful in confirming its structure in

92   P.N.W. Baxter, J.-M. Lehn, J. Fischer and M.-T. Youinou, *Angew. Chem., Int. Ed. Engl.*, 1994, **33**, 2284.

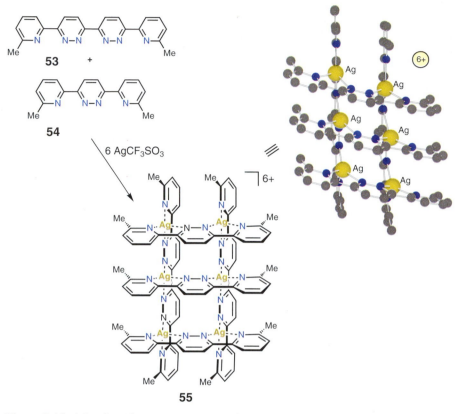

Figure 7.19 *A 3 × 2 grid incorporating six silver(I) ions*

solution. A set of 4 : 4 : 1 singlets was observed in deutero-nitromethane, corresponding to the three different environments expected for the silver ions in the product, namely the corner, side and centre sites.

It is possible to 'mix and match' ligand types to produce 'rectangular' grids. Self-assembly of the di-and tri-topic ligands **53** and **54** with silver triflate in a 2 : 3 : 6 stoichiometric ratio in nitromethane results in formation of the 2 × 3 grid **55** (Figure 7.19), with minor amounts of the corresponding 2 × 2 and 3 × 3 'square' grids also being formed.[93] The identity of the major product was established by X-ray diffraction as well as by NMR spectroscopy. The six silver centres are arranged in a rhombohedrally-distorted rectangular matrix, with the average silver–silver separation being 3.75 Å; the co-ordination geometry of each silver is distorted tetrahedral. The overall dimensions of the grid are 15.2 × 11.1 Å. It is noted that the mixed-ligand grid is formed preferentially in this system over the symmetric 2 × 2 and 3 × 3 alternatives. This preference (corresponding to hetero- over homo-

93   P.N.W. Baxter, J.-M. Lehn, B.O. Kneisel and D. Fenske, *Angew. Chem., Int. Ed. Engl.*, 1997, **36**, 1978.

ligand recognition) may arise, in part, from an avoidance of the 3 × 3 species. The latter possesses one binding site of lower binding affinity for silver(I) than the others; namely, the central tetra-pyridazine donor site (which lacks pyridyl nitrogen donors).

7.3.3 Cylindrical System

The remarkable cylindrical supramolecular structure **56** forms by spontaneous assembly from 11 particles: five ligand molecules (three 5,5′-linked bis-bipyridines and two hexa-azatriphenylene ligands) and six copper(I) ions.[94] The hexa-azatriphenylene ligands remain flat and are orientated almost parallel to each other; they are not eclipsed as shown in **56** but are rotated with respect to each other by 27°. Reflecting this, the overall structure has a triple-helical twist. After allowing for van der Waals radii, the internal cavity is about 4 Å in diameter and 4 Å high.

56

Electrospray mass spectrometry and spectrophotometric titration data have been used to obtain information about both the thermodynamics and mechanism of formation of **56**. These studies demonstrated that the assembling process occurs with positive co-operativity in this case.[95]

The spontaneous assembly of a capped trimetallic complex of type **59** from **57**, **58** and copper(I) in acetonitrile–dichloromethane has also been reported.[96] Once again the product was characterised by electrospray mass spectrometry as well as by a spectrophotometric titration study.

[94] P. Baxter, J.-M. Lehn, A. DeCian and J. Fischer, *Angew. Chem., Int. Ed. Engl.*, 1993, **32**, 69.
[95] A. Marquis-Rigault, A. Dupont-Gervais, P.N.W. Baxter, A. Van Dorsselaer and J.-M. Lehn, *Inorg. Chem.*, 1996, **35**, 2307.
[96] E. Leize, A. Van Dorsselaer, R. Kramer and J.-M. Lehn, *J. Chem. Soc., Chem. Commun.*, 1993, 990.

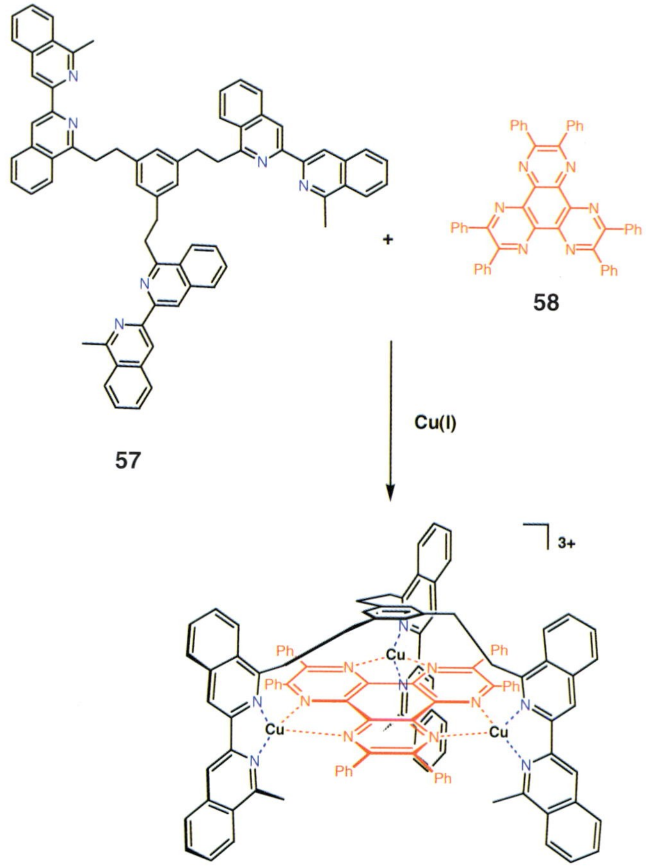

57

58

Cu(I)

59

7.3.4 Cages

Self-assembly has been employed for the construction of unique metal-containing cage structures using the procedure shown in Figure 7.20.[97] The optically active products **60** and **61** are rare examples of supramolecular systems belonging to the *T*-symmetry point group. An important element in the success of this synthesis was the use of the bis(triflato) platinum(II) and palladium(II) derivatives of the chiral ligand 2,2'-bis(diphenylphosphino)-1,1'-binaphthyl. The well documented chemistry of the platinum and palladium complexes of this ligand and, in particular, their strong affinity for pyridine nitrogens suggested their use as corner units – they had already been employed in this capacity for the assembly of optically active molecular squares.[29,98] Both **60** and **61** are microcrystalline solids with high decomposition points; they are quite robust, although hygroscopic.

[97] P.J. Stang, B. Olenyuk, D.C. Muddiman and R.D. Smith, *Organometallics*, 1997, **16**, 3094.
[98] P.J. Stang and B. Olenyuk, *Angew. Chem., Int. Ed. Engl.*, 1996, **35**, 732.

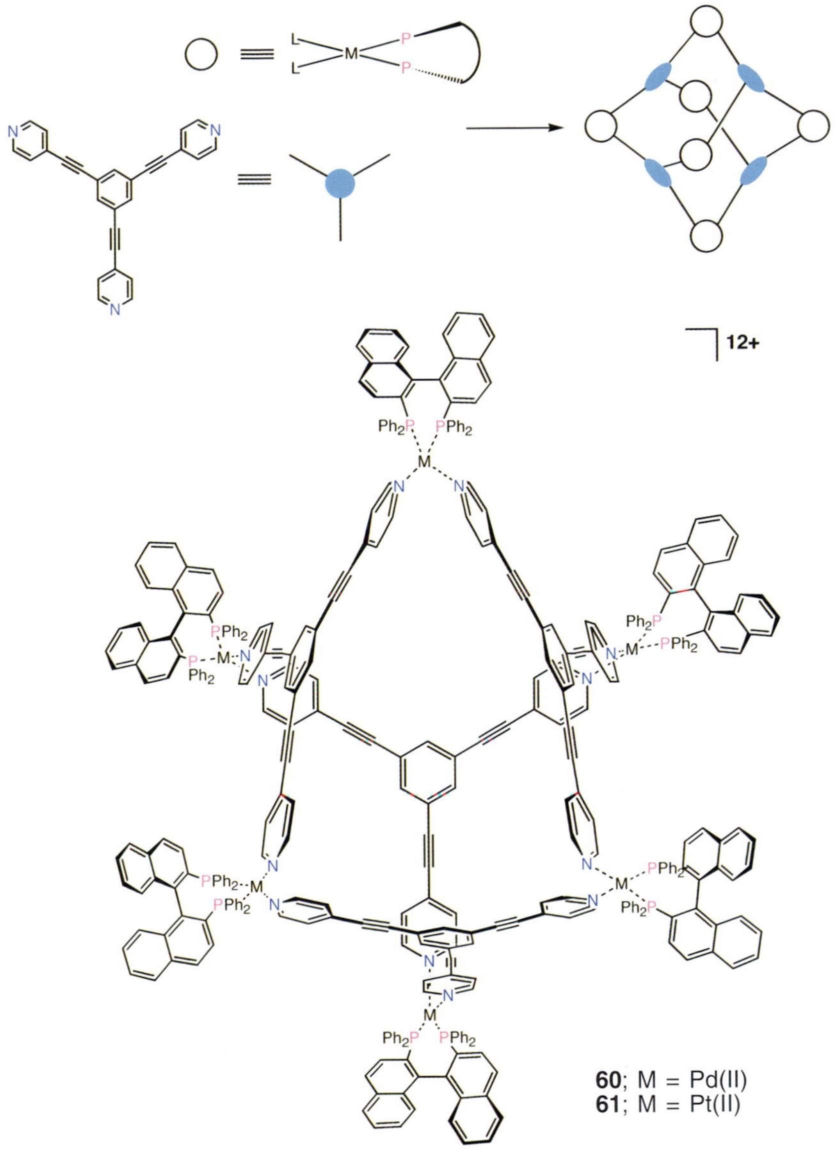

60; M = Pd(II)
61; M = Pt(II)

Figure 7.20 *Self-assembly of the large cage-like structures* **60** *and* **61**[97]

In another study, the metal-induced self-assembly of other cage stuctures, in which four platinum(II) or palladium(II) ions serve to link two preorganised (identical) cone-shaped precursors *via cis* nitrile–metal–nitrile bonds, has been reported.[99]

[99] P. Jacopozzi and E. Dalcanale, *Angew. Chem., Int. Ed. Engl.*, 1997, **36**, 613.

7.3.5 Metal-containing Receptors

Metal template self-assembly has been employed to form a number of metal-containing host assemblies which then act as specific receptors for chosen substrates.[100,101] An example of this category, involving the uptake of two glutaric acid molecules, is illustrated in Figure 7.21. In this case the selective binding of the glutaric acid guests by the copper(I) templated host was observed over other diacids of different chain length (such as pimelic and diethylmalonic acids). For particular systems, it has been demonstrated that, as expected, the choice of metal ion influences the stability of the host–guest complex formed. In particular cases, the metal is also capable of serving as a spectroscopic marker for indicating the occurrence of host–guest complexation.

In other studies, a semi-rigid linked di-zinc(II) bis-porphyrin receptor has been synthesised.[102] This Trogers base-linked system forms a molecular cleft in which the complexed zinc ions are (initially) held about 16 Å apart on either side of the cleft. Binding studies showed that 1,7-diaminoheptane is the smallest guest of this type that can bind to each zinc in a ditopic fashion. The binding constant for 1 : 1 complexation in toluene at 25 °C is 1.5×10^6 dm^3 mol^{-1}. When the related tetra-topic guest (NH$_2$CH$_2$CH$_2$CH$_2$)$_2$NCH$_2$CH$_2$CH$_2$CH$_2$N(CH$_2$CH$_2$CH$_2$NH$_2$)$_2$ was employed in a related study, a large, self-assembled spherical 'cage-like' structure, involving two molecular clefts of the above type 'encapsulating' the amine guest, formed spontaneously.

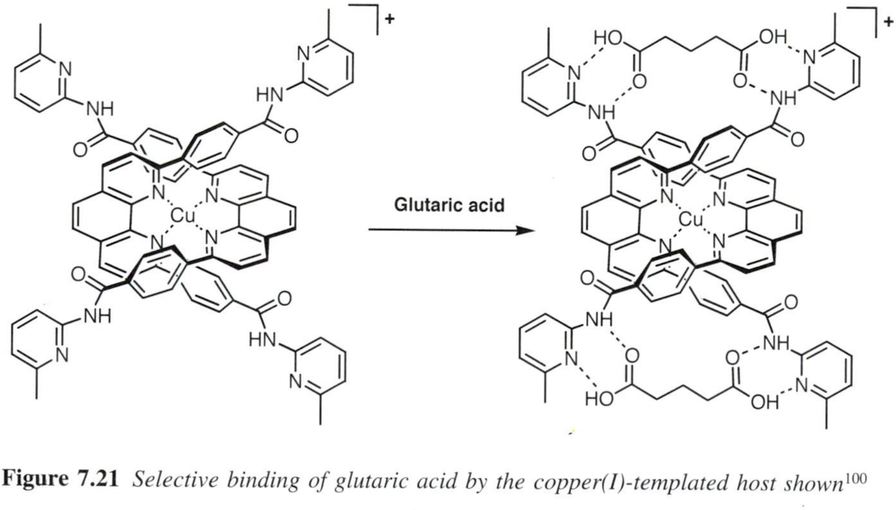

Figure 7.21 *Selective binding of glutaric acid by the copper(I)-templated host shown*[100]

[100] B. Linton and A. Hamilton, *Chemtech*, 1997 (July), 34.
[101] B. Linton and A. Hamilton, *Chem. Rev.*, 1997, **97**, 1669.
[102] J.N.H. Reed, A.P.H.J. Schenning, A.W. Bosman, E.W. Meijer and M.J. Crossley, *Chem. Commun.*, 1998, 11.

7.3.6 Metal-linked Dendrimers

A further category of (often partially) metal-directed, self-assembled supramolecules is made up of dendritic systems in which metal ions are employed as links in the radiating strands.[103,104] This category thus contrasts with the majority of reported dendrimers which employ solely metal-free covalent connectivity.[105]

The 'spherical' dendrimer **62** serves as example of the metal-linked type.[103] While both the central core (incorporating twelve terpyridine sites) and the terpyridine-containing components on the periphery of the structure are obtained by conventional organic synthesis, the overall dodeca-ruthenium structure assembles in 75% yield on reaction of the core with 15 equivalents of the 1 : 1 ruthenium(III) chloride complex of the 'terminating' terpyridyl derivative.

62

[103] G.R. Newkome, F. Cardullo, E.C. Constable, C.N. Moorefield and A.M.W.C. Thompson, *J. Chem. Soc., Chem. Commun.*, 1993, 925.

[104] D. Armspach, M. Cattalini, E.C. Constable, C.E. Housecroft and D. Phillips, *Chem. Commun.*, 1996, 1823.

[105] F. Zeng and S.C. Zimmerman, *Chem. Rev.*, 1997, **97**, 1681; O.A. Matthews, A.N. Shipway and J.F. Stoddart, *Prog. Polym. Sci.*, 1998, **23**, 1.

Subject Index